LEED v4
GREEN ASSOCIATE EXAM GUIDE (LEED GA)

Comprehensive Study Materials, Sample Questions,
Green Building LEED Certification, and Sustainability

Gang Chen

ArchiteG®, Inc.
Irvine, California

LEED v4 Green Associate Exam Guide (LEED GA):
Comprehensive Study Materials, Sample Questions, Green Building LEED Certification, and Sustainability

Copyright © 2014   Gang Chen
Edition 4.0.3   Issued on 8/8/2014

Cover Photo © 2014   Gang Chen. All rights reserved.

Copy Editor: Penny L Kortje

All Rights Reserved.
No part of this book may be transmitted or reproduced by any means or in any form, including electronic, graphic, or mechanical, without the express written consent of the publisher or author, except in the case of brief quotations in a review.

ArchiteG®, Inc.
http://www.ArchiteG.com
http://www.GreenExamEducation.com

ISBN: 978-1-61265-018-0

PRINTED IN THE UNITED STATES OF AMERICA

What others are saying about *LEED Green Associate Exam Guide*...

"Finally! A comprehensive study tool for LEED GA Prep!"
"I took the one-day Green LEED Green Associate course and walked away with a power point binder printed in very small print—which was missing MUCH of the required information (although I didn't know it at the time). I studied my little heart out and took the test, only to fail it by 1 point. Turns out I did NOT study all the material I needed to in order to pass the test. I found this book, read it, marked it up, retook the test, and passed it with a 95%. Look, we all know the LEED Green Associate Exam is new and the resources for study are VERY limited. This one's the VERY best out there right now. I highly recommend it."
—**Consultant VA**

"Complete overview for the LEED Green Associate exam"
"I studied this book for about three days and passed the exam ... if you are truly interested in learning about the LEED system and green building design, this is a great place to start."
—**K.A. Evans**

"Very effective study guide"
"I purchased both this study guide and Mr. Chen's LEED GA Mock Exams book and found them to be excellent tools for preparing for the LEED Green Associate Exam. While Mr. Chen's LEED Green Associate Exam Guide is not perfect (in that it's not the most user-friendly presentation of the material), it was very effective in at least presenting most, if not all, of the topics that the exam touched upon. While I wouldn't necessarily recommend my abbreviated strategy for preparing for the exam, the following worked for me: I read through the exam guide a couple of times (but not word for word), took the mock exam and referenced the guide for explanations for any wrong answers, did the same for the two mock exams in Mr. Chen's LEED GA Mock Exams book, flipped through the documents that Mr. Chen recommends, and took two other web-based mock exams that I purchased on eBay. Literally after ten hours of preparation time, I took the actual exam and passed with a 189, thanks in large part to Mr. Chen's books. If I decide to take one of the LEED AP exams in the future, I will definitely be picking up more of Mr. Chen's study materials."
—**shwee "shwee"**

"Only study guide needed to pass on your first try"
"I don't write reviews, but I'm compelled to for this purchase. This was the only book I read and studied to prepare for the LEED Green Associate exam, and passed with ease on the first try today. I was over prepared for the exam by using this study guide, which is what I wanted on exam day. I bought the book, read it three times, learned a lot of good information, saved valuable time, and passed on the first try. By the way, I'm not a good test taker. I don't agree with any of the negative reviews that are posted... I'm glad I ignored those when I made the purchase and went with the majority. THIS PRODUCT DELIVERED THE RESULTS, CASE CLOSED. I'll be buying his *LEED AP BD&C Exam Guide* to prepare for the specialty exam. Thank you Mr. Gang Chen!!!"
—**Lobo**

"I just finished taking the LEED Green Associate Exam and, thankfully, I passed it on the first try by using this book as my primary study guide...I particularly liked the way the author organized the information within it."
—**Lewis Colon**

"Just what is needed"
"I'm glad I used this book for my LEED Green Associate test, which I passed today. I tried other resources (stay away from the USGBC ones on their website!) but none were as organized as this book. When you're studying to pass the exam, you just need to know what to memorize (this would be different if you are trying to learn the concepts for work, which I think you're better off getting LEED project work experience!). Gang Chen's book gives you most of what you need to memorize in the core 120+ pages as opposed to other resources that are 250 pages full of fluff. Just remember, you only need to answer about 60% of the questions correct to pass the exam...

My tip is to get this book, read it three times (yes, three) to memorize the facts, take the mock test included to see where you need to improvement, and do a little outside research. You just need a 60% passing score..."
—**N. Kong**

"Best LEED Green Associate Study Package"
"This book is easy to read, and successfully gives you a much needed crash course on literally, how to study (for this exam and any other exam for that matter) and how to use the material provided most efficiently; the material in this book is able to zoom in on the most important items required to pass the exam, and deliver each item in a way which seems less technical and easier to remember (the bottom line). This book is one of a kind at the time of this review, and could be the most affordable tutor available."
—**P O "philz"**

"My name is Elizabeth (last name deleted to protect her privacy) and I am a junior at the University of....(college name deleted to protect her privacy). This summer I attained an internship with Kath Williams and Associates, a collaborative of creative independent contractors who come together to support innovative green projects, to learn about sustainable building. At the beginning of July I spoke with Kath about taking the LEED Green Associate Exam. After having no prior experience in sustainable architecture or LEED buildings I used both your review book and the study guides by USGBC as my study references. I wanted to e-mail to thank you for such a comprehensive review guide that was enormously helpful in passing the exam. Without your guide I don't know if I would have passed. Thank you so much."
—**Elizabeth**

"A Great Book for Preparing the LEED exam!"
"I have read almost all the books for LEED exams, and found LEED Exam Guide series to be the best. The USGBC Reference Guide was too detailed and kind of confusing. Some other third party books have too many grammatical mistakes, are hard to understand, and have way too many questions. The questions in those books are confusing instead of helpful. The USGBC workshop missed some of the very important information, like extra credits.
LEED Exam Guide series gives you just the right amount of information for you to pass the LEED exam. Each book in the series includes study materials, sample questions and answers, as well as mock exam and answers for a specific LEED exam. It also gives you the most information for you to get your building LEED certified. A Great book!"
—**Ellen**

"A Wonderful Guide for the LEED Green Associate Exam"
"After deciding to take the LEED Green Associate Exam, I started to look for the best possible study materials and resources. From what I thought would be a relatively easy task, it turned into a tedious

endeavor. I realized that there are vast amounts of third-party guides and handbooks. Since the official sites offer little to no help, it became clear to me that my best chance to succeed and pass this exam would be to find the most comprehensive study guide that would not only teach me the topics, but would also give me a great background and understanding of what LEED actually is. Once I stumbled upon Mr. Chen's book, all my needs were answered. This is a great study guide that will give the reader the most complete view of the LEED exam and all that it entails.

"The book is written in an easy-to-understand language and brings up great examples, tying the material to the real world. The information is presented in a coherent and logical way, which optimizes the learning process and does not go into details that will not be needed for the LEED Green Associate Exam, as many other guides do. This book stays dead on topic and keeps the reader interested in the material.

"I highly recommend this book to anyone that is considering the LEED Green Associate Exam. I learned a great deal from this guide, and I am feeling very confident about my chances for passing my upcoming exam."
—**Pavel Geystrin**

"Like other books in the LEED Exam Guide series, this is a great timesaver! The important information that you need to memorize is already highlighted / underlined by the author. This really saved me a lot of time. I love it! A Great Timesaver!"
—**Alice**

"Easy to read, easy to understand"
"I have read through the book once and found it to be the perfect study guide for me. The author does a great job of helping you get into the right frame of mind for the content of the exam. I had started by studying the Green Building Design and Construction reference guide for LEED projects produced by the USGBC. That was the wrong approach, simply too much information with very little retention. At hundred of pages in textbook format, it would have been a daunting task to get through it. Gang Chen breaks down the points, helping to minimize the amount of information but maximizing the content I was able to absorb. I plan on going through the book a few more times, and I now believe I have the right information to pass the LEED Green Associate Exam."
—**Brian Hochstein**

"All in one—LEED GA prep material"
"Since the LEED Green Associate Exam is a newer addition by USGBC, there is not much information regarding study material for this exam. When I started looking around for material, I got really confused about what material I should buy. This LEED GA guide by Gang Chen is an answer to all my worries! It is a very precise book with lots of information, like how to approach the exam, what to study and what to skip, links to online material, and tips and tricks for passing the exam. It is like the 'one stop shop' for the LEED Green Associate Exam. I think this book can also be a good reference guide for green building professionals. A must-have!"
—**SwatiD**

"I'm giving this book 5 stars, as it accomplishes exactly what the author promises. Read it multiple times and memorize the key points/stats/figures, and you will be poised to pass the test."
—**R. C. O'Brien**

"Useful Study Guide"
"The book is by far the most useful study guide I've purchased for the LEED GA exam. The other books were far too vague and did not approach the kind of pragmatic detail you'd need to pass."
—Swansong8

"Really Helpful"
"I purchased other exam preparation books for the LEED GA test and none of the others compared to this one. The layout and content of this book was much easier to use than the books from the USGBC. This book is all I needed to pass the test with its study tools like sample questions and exams. All in all a great buy!"
—Jamie

"Wonderful Resource!"
"This book is well organized and concise. The material is completely relevant and truly is what helped me to pass the LEED Green Associate Exam! I used this book as my primary study material and found it more than sufficient. I recommend this book to anyone preparing to take the LEED GA exam!"
—Joel I. Semel

"With this book, plus the little red book, *LEED Green Associate Mock Exams*, (also by Gang Chen) combined with all the free PDF downloads printed from the USGBC website, I passed my LEED GA exam on the first try! Gang Chen's books provide a concise and organized framework of all the info you need...Highly recommended."
—Sarah Bartz

"Excellent resource!!!"
"I reviewed this book and memorized the charts and passed on the first try! This along with the practice exams by Gang Chen were all that I needed and included valuable tips for studying and additional (free) publications to refer to. This was an excellent purchase."
—K. Lairet

"Great Study guide. Will DEFINITELY help you pass the exam!"
"I took the LEED Green Associate Exam this morning and passed with a score of 190/200. I used this book as my primary study tool. The only other thing I did was to read the Green Associates handbook and I used the LEED NC Reference Guide to check the information in the book. The book is broken down very well for studying making it much easier to memorize the data. The practice exam will prepare you well. It is much harder than the real exam. There was a lot of information in this book that showed up on the exam that I would not have known had I not read it.

My basic study schedule was about one day of reading the nontechnical chapters, one day devoted to the technical chapter about credits, and then the morning of the test to review my highlighted notes. I think if anyone studies this book until they pass the mock exam (I scored about a 70/100) and feel comfortable then they will be over prepared for the actual exam."
—K. Mitchell "Kmitchell"

Leadership in Energy and Environmental Design (LEED)

LEED-CERTIFIED   LEED-SILVER
LEED-GOLD   LEED-PLATINUM

LEED GREEN ASSOCIATE

LEED AP BD+C   LEED AP ID+C
LEED AP O+M
LEED AP HOMES   LEED AP ND

LEED FELLOW

Dedication

To my parents, Zhuixian and Yugen,
my wife Xiaojie, and my daughters
Alice, Angela, Amy, and Athena.

Disclaimer

This book provides general information about the LEED Green Associate Exam and green building LEED Certification. It is sold with the understanding that the publisher and author are not providing legal, accounting, or other professional services. If legal, accounting, and other professional services are required, seek the services of a competent professional firm.

It is not the purpose of this book to reprint the content of all other available texts on the subject. You are urged to read other available texts and tailor them to fit your needs.

Great effort has been made to make this book as complete and accurate as possible; however, nobody is perfect, and there may be typographical or other mistakes. You should use this book as a general guide and not as the ultimate source on this subject. If you find any potential errors, please send an e-mail to: info@ArchiteG.com

This book is intended to provide general, entertaining, informative, educational, and enlightening content. Neither the publisher nor the author shall be liable to anyone or any entity for any loss or damages, or alleged loss and damages, caused directly or indirectly by the content of this book.

USGBC and LEED are trademarks of the US Green Building Council. The US Green Building Council is not affiliated with the publication of this book.

If you do not wish to be bound by the above, you may return this book to the publisher for a full refund.

Legal Notice

Contents

Preface

Chapter One   LEED Exam Preparation Strategies, Methods, Tips, Suggestions, Mnemonics, and Exam Tactics to Improve Your Exam Performance..................................23

1. The nature of LEED exams and exam strategies...23
2. LEED exam preparation requires short-term memory..24
3. LEED exam preparation strategies and scheduling...24
4. Timing of review: the 3016 rule; memorization methods, tips, suggestions, and mnemonics........25
5. The importance of good and effective study methods..27
6. The importance of repetition: read this book at least three times.......................................27
7. When should you start to do sample questions and mock exams?......................................28
8. How much time do you need for LEED exam preparation?...28
9. The importance of a routine..28
10. The importance of short, frequent breaks and physical exercise..28
11. A strong vision and a clear goal...29

Chapter Two   Overview..31

1. What is LEED? What is the difference between LEED, LEED AP, LEED Green Associate, LEED AP+, and LEED Fellow?..31
2. Why did the GBCI create the three-tier LEED credential system?....................................33
3. Do I need to have LEED project experience to take the LEED exams?............................33
4. How do I become a LEED AP+? Do I have to take the LEED Green Associate Exam first to become a LEED AP+?..33
5. How many questions do you need to answer correctly to pass the LEED exams?............34
6. What are the key areas that USGBC uses to measure the performance of a building's sustainability?...35
7. How many LEED exams does USGBC have?...35
8. Are the LEED exams valid and reliable?...35
9. How many member organizations does the USGBC have?...35
10. How many regional chapters does the USGBC have?...35
11. What is the main purpose of the USGBC?...35
12. What are the guiding principles of the USGBC?..36
13. How much energy and resources do buildings consume in the US?..................................36
14. What is the most important step to get your building certified?..36
15. What are the benefits of green buildings and LEED Certification?...................................36
16. Who developed the LEED green building rating systems?..36
17. What current reference guides and specific green building rating systems does the USGBC have?..36
18. How does LEED fit into the green building market?...37
19. What are the benefits of LEED certification for your building?..37

20.	What is the procedure of LEED certification for your building?................................	38
21.	How much is the building registration fee and how much is the building LEED certification fee?..	38
22.	What are LEED's system goals?..	38
23.	How are LEED credits allocated and weighted?...	38
24.	Are there LEED certified products?..	39

Chapter Three Introduction to the LEED Green Associate Exam............................41

1.	What is new for LEED v4?...	41
2.	What is the scope of the LEED Green Associate Exam?......................................	41
3.	What is the latest version of LEED and when was it published?...........................	43
4.	How many possible points does LEED v4 have?..	44
5.	How many different levels of building certification does USGBC have?...............	45
6.	What is the process or basic steps for LEED certification?..................................	45
7.	What does the registration form include?...	48
8.	What is precertification?..	48
9.	What is a CIR?...	48
10.	When do you submit a CIR?..	48
11.	What are the steps for submitting a CIR?...	49
12.	Will a CIR guarantee a credit?...	49
13.	What tasks are handled by GBCI and what tasks are handled by USGBC?........	49
14.	What are MPRs?..	49
15.	Which LEED rating system should I choose?...	50

Chapter Four LEED Green Associate Exam Technical Review................................51

1.	What do green buildings address?..	51
2.	Key stake holders and an integrative approach..	51
3.	The mission of USGBC..	51
4.	The structure of LEED rating system...	52
5.	LEED certification tools...	52
6.	**Specific Technical Information**...	53

Chapter Five Integrative Process (IP)..55

*IP Prerequisite (IPP): Integrative Project Planning and Design................................... 55
IP Credit (IPC): Integrative Process.. 56

Note: Each credit is described in a standard format, including Purpose, Concept, and Synergies.

A few of the credits are for one or two of the rating systems only, and are <u>unlikely to be tested on the LEED Green Associate Exam</u>. We just list these credits' names and related points, omitting their detailed discussions.

| Chapter Six | Location and Transportation (LT) | 59 |

Overall Purpose...59
Consistent Documentation..59
LT Credit (LTC): LEED for Neighborhood Development Location.............................60
LTC: Sensitive Land Protection..61
LTC: High-Priority Site...62
LTC: Surrounding Density and Diverse Uses...63
LTC: Access to Quality Transit..64
LTC: Bicycle Facilities...65
LTC: Reduced Parking Footprint..66
LTC: Green Vehicles..67

| Chapter Seven | Sustainable Sites (SS) | 69 |

Overall Purpose...69
Core Concepts..70
Recognition, Regulation, and Incentives...70
Overall Strategies and Technologies...70
SS Prerequisite (SSP): Construction Activity Pollution Prevention............................71
SSP: Environmental Site Assessment..72
SSC: Site Assessment..73
SSC: Site Development—Protect or Restore Habitat..74
SSC: Open Space...75
SSC: Rainwater Management..76
SSC: Heat Island Reduction..77
SSC: Light Pollution Reduction..78
*SSC: Site Master Plan..79
*SSC: Tenant Design and Construction Guidelines...79
*SSC: Places of Respite..79
*SSC: Direct Exterior Access...79
*SSC: Joint Use of Facilities..79

| Chapter Eight | Water Efficiency (WE) | 81 |

Overall Purpose...81
Cross-Cutting Issues..81
Core Concepts..82
Recognition, Regulation, and Incentives...82
Overall Strategies and Technologies...83
WE Prerequisite (WEP): Outdoor Water Use Reduction..85
WEP: Indoor Water Use Reduction...86
WEP: Building-Level Water Metering..87
WEC: Outdoor Water Use Reduction..88
WEC: Indoor Water Use Reduction...89
WEC: Cooling Tower Water Use...90
WEC: Water Metering..91

Chapter Nine   Energy and Atmosphere (EA)..95

Overall Purpose..95
Core Concepts...96
Recognition, Regulation, and Incentives...96
Overall Strategies and Technologies...96
EA Prerequisite (EAP): Fundamental Commissioning and Verification.................98
EAP: Minimum Energy Performance...99
EAP: Building-Level Energy Metering..100
EAP: Fundamental Refrigerant Management...101
EAC: Enhanced Commissioning..102
EAC: Optimize Energy Performance..103
EAC: Advanced Energy Metering..104
EAC: Demand Response...105
EAC: Renewable Energy Production..106
EAC: Enhanced Refrigerant Management..107
EAC: Green Power and Carbon Offsets..108

Chapter Ten   Materials and Resources (MR)..111

Overall Purpose..111
The Hierarchy of Waste Management...111
Life-Cycle Assessment in LEED..112
Cross-Cutting Issues...112
Core Concepts..115
Recognition, Regulation, and Incentives..116
Overall Strategies and Technologies..116
MR Prerequisite (MRP): Storage and Collection of Recyclables............................117
MRP: Construction and Demolition Waste Management Planning........................118
*MRP: PBT Source Reduction—Mercury...119
MRC: Building Life-Cycle Impact Reduction...120
MRC: Building Product Disclosure and Optimization—Environmental Product Declarations............122
MRC: Building Product Disclosure and Optimization—Sourcing of Raw Materials..................123
MRC: Building Product Disclosure and Optimization—Material Ingredients........................124
MRC: PBT Source Reduction—Mercury..127
MRC: PBT Source Reduction—Lead, Cadmium, and Copper...............................127
MRC: Furniture and Medical Furnishings..127
MRC: Design for Flexibility...127
MRC: Demolition Waste Management..128

Chapter Eleven   Indoor Environmental Quality (EQ)..129

Overall Purpose..129
Cross-Cutting Issues...129
Core Concepts..132
Recognition, Regulation, and Incentives..133
Overall Strategies and Technologies..133

EQP: Minimum Indoor Air Quality Performance..134
EQP: Environmental Tobacco Smoke Control..135
EQP: Minimum Acoustic Performance...136
EQC: Enhanced Indoor Air Quality Strategies..137
EQC: Low-Emitting Materials..138
EQC: Construction Indoor Air Quality Management Plan..139
EQC: Indoor Air Quality Assessment...140
EQC: Thermal Comfort...141
EQC: Interior Lighting..142
EQC: Daylight...143
EQC: Quality Views..144
EQC: Acoustic Performance...145

Chapter Twelve   Innovation (IN)..147

Overall Purpose..147
Innovation Credit (INC): Innovation..148
INC: LEED Accredited Professional..149

Chapter Thirteen   Regional Priority (RP)..151

Overall Purpose..151
Regional Priority Credit (RPC): Regional Priority..152

Chapter Fourteen   LEED Green Associate Exam Samples Questions, Answers, and Exam Registration..153

I.   LEED Green Associate Exam sample questions...153
II. Answers for the LEED Green Associate Exam sample questions..............................158
III. Where can I find the latest official sample questions for LEED Green Associate Exam?.............163
IV. LEED Green Associate Exam registration..163

Chapter Fifteen   Frequently Asked Questions (FAQ) and Other Useful Resources..............165

1. I found the reference guide way too tedious. Can I only read your book and just refer to the USGBC reference guide (if one is available for the exam I am taking) when needed?................165
2. Is one week really enough for me to prepare for the exam while I am working?..........165
3. Will your book(s) be adequate for me to pass the exam?..165
4. I am preparing for the LEED exam. Do I need to read the 2" thick reference guide?....166
5. For LEED v4, will the total number of points be more than 110 in total if a project gets all of the extra credits and all of the standard credits?...166
6. Are you writing new versions of books for the new LEED exams? What new books are you writing?..166
7. Important documents that you need to download for free, become familiar with, and memorize......166
8. Important documents that you need to download for free, and become familiar with....167

9. Do I need to take many practice questions to prepare for a LEED exam?...........................168

Appendixes..171
1. Default occupancy factors..171
2. Important resources and further study materials you can download for free................171
3. Annotated bibliography..172
4. Valuable websites and links...172
5. Important items covered by the second edition of *Green Building and LEED Core Concepts Guide*.173

Back page promotion..177
1. LEED Exam Guide series (ArchiteG.com)..179
2. *Building Construction*..182
3. *Planting Design Illustrated*..183

Index...185

Preface

The USGBC released LEED v4 at the GreenBuild International Conference and Expo in November 2013. The GBCI started to include the new LEED v4 content for all LEED exams in late Spring 2014. We have incorporated the new LEED v4 content in this book.

Starting on December 1, 2011, GBCI began to draw LEED Green Associate Exam questions from the second edition of *Green Building and LEED Core Concepts Guide*. We have incorporated this latest information in our book. See appendix 5 for "Important items covered by the second edition of *Green Building and LEED Core Concepts Guide*."

The main purposes of this book are to help you pass the LEED Green Associate Exam and to assist you with understanding the process of getting your building LEED certified.

The LEED Green Associate Exam is the most important LEED exam for two reasons:

1. You have to pass it in order to get the title of LEED Green Associate.

2. The exam is the required Part One (2 hours) of ALL LEED AP+ exams. You have to pass it before taking Part Two (2 hours) of the specific LEED AP+ exam of your choice to get any LEED AP+ title unless you have passed the old LEED AP exam before June 30, 2009.

There are a few ways to prepare for the LEED Green Associate Exam:

1. Take USGBC courses or workshops. You should take USGBC classes at both the 100 (Awareness) and 200 (LEED Core Concepts and Strategies) levels to successfully prepare for the exam. A one-day course can cost $445 if you register early enough, and can be as expensive as $495 if you miss the early bird special. You will also have to wait until the USGBC workshops or courses are offered in a city near you.

OR
2. Take USGBC online courses. You can go to the USGBC or GBCI websites for information. The USGBC online courses are less personal and still expensive.

OR
3. Read related books. GBCI is now handling the LEED exams, instead of USGBC. Unfortunately, there is NO official GBCI book on the LEED Green Associate Exam. However, there are a few third party books on the LEED Green Associate Exam. *LEED Green Associate Exam Guide (LEED GA)* is one of the first books covering the current version of the exam and will fill in this blank to assist you with passing the exam.

To stay at the forefront of the LEED and green building movement and make my books more valuable to their readers, I sign up for USGBC courses and workshops myself, and I review the USGBC and GBCI websites and many other sources to get as much information as possible on LEED. *LEED Green Associate Exam Guide (LEED GA)* is a result of this very comprehensive research. I have done the hard work so that you can save time preparing for the exam by reading my book.

Strategy 101 for the LEED Green Associate Exam is that you must recognize that you have only a limited amount of time to prepare for the exam. So, you must concentrate your time and effort on the most important content of the LEED Green Associate Exam. To assist you with achieving this goal, the book is broken into two major sections: (1) the study materials and (2) the sample questions and mock exam.

Chapter 1 covers LEED exam preparation strategies, methods, tips, suggestions, mnemonics, and exam tactics to improve your exam performance.

Chapters 2 and 3 cover general information. I use the question and answer format to try to give you the most comprehensive coverage on the subject of the LEED Green Associate Exam. I have given you only the correct answers and information so you do not need to waste your time reading and remembering the wrong information. As long as you understand and remember the correct information, you can pass the test, no matter how the USGBC changes the format of the exam.

Chapter 4 contains a LEED Green Associate Exam technical review including overall purpose, mnemonics, core concepts, recognition, regulation, incentives, overall strategies and technologies, and some **specific technical information**.

Specific technical information for **each credit** includes purpose, credit concept, and synergies.

A few of the credits are only for one or two of the specific rating systems, and are <u>unlikely to be tested on the LEED Green Associate Exam</u>. We just list these credit names and related points, omitting their detailed discussions.

In the back section, you will find sample questions. These are intended to match the latest real LEED Green Associate Exam as closely as possible and assist you in becoming familiar with the format of the exam.

Most people already have some knowledge of LEED. I suggest that you use a highlighter when you read this book; you can highlight the content that you are not familiar with when you read the book for the first time. You can try to cover the answer and read a question first. If you can come up with the correct answer before you read the book, you do not need to highlight the question and answer. If you cannot come up with the correct answer before you read the book, then highlight that question. This way, when you do the review later and read the book for the second time, you can just focus on the portions that you are not familiar with and save yourself a lot of time. You can repeat this process with different colored highlighters until you are very familiar with the content of this book. Then, you will be ready to take the LEED Green Associate Exam.

The key to passing the LEED Green Associate Exam, or any other exam, is to know the scope of the exam, and not to read too many books. Select one or two really good books and focus on them. Actually <u>understand</u> the content and <u>memorize</u> it. For your convenience, I have <u>underlined</u> the fundamental information that I think is very important. You definitely need to <u>memorize</u> all the information that I have underlined. You should try to understand the concept first, and then memorize the content of the book by reading it multiple times. This is a much better way to prepare, rather than "mechanical" memory without understanding.

The part of the LEED Green Associate Exam that you can control through preparation and simple memorization is the section regarding the credit concept and credit process for the LEED building rating system. You should become very familiar with every major credit category. You should try to answer all questions related to this part correctly.

There is also a part of the exam that you may not be able to control. You may not have experience preparing actual LEED building certification, so there will be some questions that may require you to guess. This is the hardest part of the exam, but these questions should be only a small percentage of the test if you are well prepared. You should <u>eliminate</u> the obvious wrong answers and then attempt an educated <u>guess</u>. There is no penalty for guessing. If you have no idea what the correct answer is and cannot eliminate any obvious wrong answer, then do not waste too much time on the question, just pick a guess answer. The key is to try and use the <u>same</u> guess answer for all of the questions that you have no idea about. For example, if you choose "a" as the guess answer, then you should be consistent and use "a" as the guess answer for all the questions that you have no idea about. That way, you will likely have a better chance at guessing more correct answers.

The actual LEED Green Associate Exam has 100 multiple-choice questions and you must finish it within two hours. The raw exam score is converted to a scaled score ranging from 125 to 200. The passing score is 170 or higher. You need to answer <u>about</u> 60 questions correctly to pass. There is an optional ten-minute tutorial for computer testing before the exam and an optional ten-minute exit survey.

This is not an easy exam, but you should be able to pass it if you prepare well. If you <u>set your goal for a high score and study hard</u>, you will have a better chance of passing. If you set your goal for the minimum passing score of 170, you will probably end up scoring 169 and fail, and you will have to <u>retake</u> the exam again. That will be the last thing you want. Give yourself plenty of time and do not wait until the last minute to begin preparing for the exam. I have met people who have spent 40 hours preparing and passed the exam, but I suggest that you give yourself <u>at least two to three weeks</u> of preparation time. On the night before the exam, you should look through the questions on the mock exam that you did not answer correctly and remember what the correct answers are. Read this book carefully, prepare well, relax and put yourself in the best physical and psychological state on the day of the exam, and you will pass.

Chapter 1
LEED Exam Preparation Strategies, Methods, Tips, Suggestions, Mnemonics, and Exam Tactics to Improve Your Exam Performance

1. **The nature of LEED exams and exam strategies**
LEED exams are standardized tests. They should be consistent and legally defensible tests.

The earliest standardized tests were the Imperial Examinations in China that started in 587. In Europe, traditional school exams were oral exam, and the first European written exams were held at Cambridge University, England in 1792.

Most exams test knowledge, skills, and aptitudes. There are several types of exams:
1) math exams testing your abilities to do various calculations
2) analytical exams testing skills in separating a whole (intellectual or substantial) into its elemental parts or basic principles
3) knowledge exams testing your expertise, skills, and understanding of a subject
4) creativity exams testing your skills to generate new ideas or concepts
5) performance exams such as driving tests or singing competitions

LEED exams test your knowledge of LEED as well as some very basic analytical and mathematical calculation skills (on average 10% to 15% of the exam questions require math calculations). The LEED Green Associate Exam is more like a history or political science test, which requires a lot of memorization of LEED information and knowledge.

All LEED exams test candidates' abilities at three hierarchical cognitive levels:
1) **recognition** (ability to recall facts)
2) **application** (ability to use familiar principles or procedures to solve a new problem)
3) **analysis** (ability to break the problem down into its parts, to evaluate their interactions or relationship, and to create a solution)

A LEED exam has 100 multiple-choice questions. You need to pick <u>one, two, three, or even four correct statements</u> (some questions have five <u>statements</u>), depending on the specific question.

The exam writers usually use errors that people are likely to make to create the incorrect choices (or **distracters**) to confuse exam takers, so you **HAVE to read the question very carefully**, pay special attention to words like <u>may</u>, <u>might</u>, <u>could</u>, etc. Creating effective <u>distracters</u> is key to creating a good exam. This means that the more confusing a question is, the easier it is for the GBCI to separate candidates with strong LEED knowledge from the ones with weak LEED knowledge.

Since most LEED exam questions have four choices, the **distracters** in a strong LEED exam should be able to attract at least 25% of the weakest candidates to reduce the effectiveness of guessing.

For exam takers, it is to your advantage to guess for questions that you do not know, or are not sure about, because the exam writers expect you to guess. They are trying to mislead you with the questions to make your guessing less effective. If you do not guess and do not answer the questions, you will be at a disadvantage when compared with other candidates. Eliminate the obvious wrong answers and then try an educated guess. It is better to guess with the same letter answer for all the questions that you do not know, unless it is an obvious wrong answer and has been eliminated.

2. **LEED exam preparation requires short-term memory**
   Now that you know the nature of the LEED exam, you should understand that LEED exam preparation requires **short-term memory**. You should schedule your time accordingly. In the early stages of your LEED exam preparation, you should focus on understanding and an **initial** review of the material. In the late stages of your exam preparation, you should focus on memorizing the material as a **final** review.

3. **LEED exam preparation strategies and scheduling**
   You should spend about 60% of your effort on the most important and fundamental LEED material, about 30% of your effort on sample exams, and the remaining 10% on improving your weakest areas, i.e., reading and reviewing the questions that you answered incorrectly, reinforcing the portions that you have a hard time memorizing, etc.

   Do NOT spend too much time looking for obscure LEED information because the GBCI will HAVE to test you on the most **common** LEED information. At least 80% to 90% of the LEED exam content will have to be the most common, important, and fundamental LEED knowledge. The exam writers can word their questions to be tricky or confusing, but they have to limit themselves to the important content; otherwise, their tests will NOT be legally defensible. At most, 10% of their test content can be obscure information. You only need to answer about 60% of all the questions correctly. So, if you master the common LEED knowledge (applicable to 90% of the questions) and use the guess technique for the remaining 10% of the questions on the obscure LEED content, you will do well and pass the exam.

   On the other hand, if you focus on the obscure LEED knowledge, you may answer the entire 10% obscure portion of the exam correctly, but only answer half of the remaining 90% of the common LEED knowledge questions correctly, and you will fail the exam. That is why we have seen many smart people who can answer very difficult LEED questions correctly fail. They are either use to being able to look up the answers in a book or are experienced in certain obscure items through previous quality research, but end up failing the LEED exam because they cannot memorize the common LEED knowledge needed on the day of the exam. The LEED exam is NOT an open-book exam, and you cannot look up information during the exam.

   The **process of memorization** is like **filling a cup with a hole at the bottom**. You need to fill it faster than the water leaks out at the bottom, and you need to constantly fill it; otherwise, it will quickly be empty.

   Once you memorize something, your brain has already started the process of forgetting it. It is natural. That is how we have enough space left in our brain to remember the really important things.

   It is tough to fight against your brain's natural tendency to forget things. Acknowledging this truth and the fact that you cannot memorize everything you read, you need to focus your limited time, energy, and brain power on the most important issues.

The biggest danger for most people is that they memorize the information in the early stages of their LEED exam preparation, but forget it before or on the day of the exam and still THINK they remember them.

Most people fail the exam NOT because they cannot answer the few "advanced" questions on the exam, but because they canNOT recall the information they have read on the day of the exam. They spend too much time preparing for the exam, drag the preparation process on too long, seek too much information, go to too many websites, do too many practice questions or too many mock exams (one or two sets of mock exams can be good for you), and **spread themselves too thin**. They end up **missing out on the most important information** of the LEED exam, and they fail.

The LEED Exam Guide series along with the tips and methodology in each of these books will help you MEMORIZE the most important aspects of the test to pass the exam ON THE FIRST TRY.

So, if you have a lot of time to prepare for the LEED exam, you should plan your effort accordingly. You want your LEED knowledge to peak at the time of the exam, not before or after.

For example, if you have two months to prepare for the exam, you may want to spend the first month focused on reading and understanding all of the study materials you can find as your **initial** review. Also during this first month, you can start memorizing after you understand the materials as long as you know you HAVE to review the materials again later to retain the information. If you have memorized something once, it is easier to memorize it again later.

Next, you can spend two weeks focused on memorizing the material. You need to review the material at least three times. You can then spend one week on mock exams. The last week before the exam, focus on retaining your knowledge and reinforcing your weakest areas. Read the mistakes that you have made and think about how to avoid them during the real exam. Set aside a mock exam that you have not taken and take it seven days before test day. This will alert you to your weaknesses and provide direction for the remainder of your studies.

If you have one week to prepare for the exam, you can spend two days reading and understanding the study material, two days repeating and memorizing the material, two days on mock exams, and one day retaining the knowledge and enforcing your weakest areas.

The last one to two weeks before the LEED exam is absolutely critical. You need to have the "do or die" mentality and be ready to study hard to pass the exam on your first try. That is how some people are able to pass the LEED exam with only one week of preparation.

4. **Timing of review: the 3016 rule; memorization methods, tips, suggestions, and mnemonics**
Another important strategy is to review the material in a timely manner. Some people say that the best time to review material is between 30 minutes and 16 hours (the **3016** rule) after you read it for the first time. So, if you review the material right after you read it for the first time, the review may not be helpful.

I have personally found this method extremely beneficial. The best way for me to memorize study materials is to review what I learn during the day again in the evening. This, of course, happens to fall within the timing range mentioned above.

Now that you know the **3016** rule, you may want to schedule your review accordingly. For example, you may want to read new study materials in the morning and afternoon, then after dinner do an initial review of what you learned during the day.

OR

If you are working full time, you can read new study materials in the evening or at night and then get up early the next morning to spend one or two hours on an initial review of what you learned the night before.

The initial review and memorization will make your final review and memorization much easier.

Mnemonics are **also** a very good way for you to memorize facts and data that are otherwise very hard to memorize. They are often arbitrary or illogical but they work.

A good mnemonic can help you remember something for a long time or even a lifetime after reading it just once. Without the mnemonics, you may read the same thing many times and still not be able to memorize it.

There are several common types of mnemonics:
1) **Visual** mnemonics link what you want to memorize to a visual image.
2) **Spatial** mnemonics link what you want to memorize to a space, and the order of things in it.
3) **Group** mnemonics break up a difficult piece into several smaller and more manageable groups or sets, and you memorize the sets and their order. One example is the grouping of the US 10 digit phone number into three groups. This makes the number much easier to memorize.
4) **Architectural** mnemonics are a combination of visual mnemonics, spatial mnemonics, and group mnemonics.

Imagine you are walking through a building several times, along the same path. You should be able to remember the order of each room. You can then break up the information that you want to remember and link this information to several images, and then imagine that you hang the images on walls of various rooms. You should be able to easily recall each group in an orderly manner by imagining you are walking through the building again on the same path, and looking at the images hanging on walls of each room. When you look at the images on the wall, you can easily recall the related information.

You can use your home, office, or another building that you are familiar with to build an architectural mnemonic to help you to organize the things you need to memorize.

5) **Association** mnemonics allow you to associate what you want to memorize with a sentence, a similarly pronounced word, or a place you are familiar with, etc.
6) **Emotion** mnemonics use emotion to fix an image in your memory.
7) **First letter** mnemonics use the first letter of what you want to memorize to construct a sentence or acronym. For example, "**Roy G. Biv**" can be used to memorize the order of the 7 colors of the rainbow; it is composed of the first letter of each primary color.

You can use association mnemonics to memorize all the plumbing fixtures by thinking of the typical fixtures found in your own home, PLUS a urinal.

OR
You can use "Water S K U L" (**first letter** mnemonics) to memorize them.

    <u>W</u>ater Closet
    <u>S</u>hower
    <u>K</u>itchen Sink
    <u>U</u>rinal
    <u>L</u>avatory

Another example of **first letter** mnemonics to memorize the mandatory materials that can be collected for recycling for the LEED Materials and Resources (MR) credit category is as follows:

"<u>P</u>eople <u>C</u>an <u>M</u>ake <u>G</u>reen <u>P</u>romises"

    <u>P</u>aper
    <u>C</u>ardboard
    <u>M</u>etal
    <u>G</u>lass
    <u>P</u>lastics

5. **The importance of good and effective study methods**
   There is a saying, "Give a man a fish and you feed him for a day. Teach a man to fish and you feed him for a lifetime." I think there is some truth to this. Similarly, it is better to teach someone HOW to study than just give him good study materials. In this book, I give you good study materials to save you time, but more importantly, I want to teach you effective study methods so you can not only study and pass the LEED exams, but also benefit throughout the rest of your life for anything else you need to study or achieve. For example, I give you examples of mnemonics, but I also teach you the more important lesson: HOW to make mnemonics.

   Often in the same class, all the students study almost the SAME materials, but there are some students that always manage to stay at the top of the class and get good grades on exams. Why? One very important factor is they have good study methods.

   Hard work is important, but it needs to be combined with effective study methods. I think people need to work hard AND work SMART to be successful at their work, career, or anything else they are pursuing.

6. **The importance of repetition: read this book <u>at least</u> three times**
   Repetition is one of the most important tips for learning. That is why I have listed it under a separate title. For example, you should treat this book as the core study material for your LEED exam and you need to read this book <u>at least three times</u> to get all of its benefits:

   1) The first time you read the book, everything may be new information. You should focus on understanding and digesting the materials, and also do an <u>initial</u> review with the **3016** rule. Highlight any information that you have difficulty understanding.
   2) On the second read, focus on the parts <u>I</u> have already highlighted AND the parts <u>you</u> have <u>highlighted</u> (the important parts and the weakest parts for you).

3) The third time, focus on <u>memorizing</u> the information. Remember the analogy of the <u>memorization process</u> as **filling a cup with a hole on the bottom**? Do NOT stop reading this book until you pass the real exam.

7. **When should you start to do sample tests and mock exams?**
   After reading the study materials in this book at least three times, you can start to do mock exams.

8. **How much time do you need for LEED exam preparation?**
   Do not give yourself too much time to prepare for the LEED exam. Two months is probably the maximum time you should allow for preparing for the LEED exam.

   Do not give yourself too little time to prepare for LEED exam. You want to force yourself to focus on the LEED exam but you do NOT want to give yourself too little time and fail the exam. One week is probably the minimum time you should allow for preparing for the LEED exam.

9. **The importance of a routine**
   A routine is very important for studying. You should try to set up a routine that works for you. First, look at how much time you have to prepare for the LEED exam, and then adjust your current routine to include LEED exam preparation. Once you set up the routine, stick with it.

   For example, you can spend from 8:00 a.m. to 12:00 noon, and 1:00 p.m. to 5:00 p.m. on studying new LEED exam study materials, and 7:00 p.m. to 10:00 p.m. to do an initial review of what you learned during the daytime. Then, switch your study content to mock exams, memorization and retention when it gets close to the exam date. This way, you have 11 hours for LEED exam preparation everyday. You can probably pass the LEED exam in one week with this method. Just keep repeating it as a way to <u>retain</u> the LEED knowledge.

   OR
   You can spend 7:00 p.m. to 10:00 p.m. on studying new LEED exam study materials, and 6:00 a.m. to 7:00 a.m. to do an initial review of what you learned the evening before. This way, you have four hours for LEED exam preparation every day. You can probably pass LEED in five weeks with this LEED preparation schedule.

   A routine can help you to memorize important information because it makes it easier for you to concentrate and work with your body clock.

   Do NOT become panicked and change your routine as the exam date gets closer. It will not help to change your routine and pull all-nighters right before the exam. In fact, if you pull an all-nighter the night before the exam, you may do much worse than you would have done if you kept your routine. All-nighters or staying up late are not effective. For example, if you break your routine and stay up one-hour late, you will feel tired the next day. You may even have to sleep a few more hours the next day, adversely affecting your study regimen.

10. **The importance of short, frequent breaks and physical exercise**
    Short, frequent breaks and physical exercise are VERY important for you, especially when you are spending a lot of time studying. They help relax your body and mind, making it much easier for you to concentrate when you study. They make you more efficient.

Take a five-minute break, such as a walk, at least once every one to two hours. Do at least 30 minutes of physical exercise every day.

If you feel tired and cannot concentrate, stop, go outside, and take a five-minute walk. You will feel much better when you come back.

You need your body and brain to work well to be effective with your studying. Take good care of them. You need them to be well-maintained and in excellent condition. You need to be able to count on them when you need them.

If you do not feel like studying, maybe you can start a little bit on your studies. Just casually read a few pages. Very soon, your body and mind will warm up and you will get into study mode.

Find a room where you will NOT be disturbed when you study. A good study environment is essential for concentration.

11. A strong vision and a clear goal

You need to have a strong vision and a clear goal to master the LEED system and become very familiar with the LEED certification process. This is your number one priority. You need to master the LEED knowledge BEFORE you do any sample questions or mock exams. It will make the process much easier. Otherwise, there is nothing in your brain to be tested. Everything we discuss in this chapter is to achieve this goal.

As I have mentioned on many occasions, and I say it one more time here because it is so important:

It is how much LEED information you can understand, digest, memorize, and firmly retain that matters, not how many books you read or how many sample tests you have taken. The books and sample tests will NOT help you if you cannot understand, digest, memorize, and retain the important information for the LEED exam. Cherish your limited time and effort and focus on the most important information.

Chapter 2
Overview

1. **What is LEED? What is the difference between LEED, LEED AP, LEED Green Associate, LEED AP+, and LEED Fellow?**
   Answer: LEED is a term for <u>buildings</u>. It stands for the Leadership in Energy and Environmental Design (LEED) Green Building Rating System™. It is a voluntary system set up by the US Green Building Council (USGBC) to measure the sustainability and performance of a building.

   LEED AP, LEED Green Associate, LEED AP+, and LEED Fellow are terms for <u>people</u>.

   LEED AP stands for LEED Accredited Professionals. A LEED AP is a person who has passed at least one of the three <u>old</u> versions of the LEED exams (<u>LEED-NC, LEED-CI, and LEED-EB</u>) before June 30, 2009 and has the skills and knowledge to encourage and support integrated design, to take part in the design process, and to control the application and certification process for a LEED building.

   LEED APs have **three choices:**

   1) <u>Do nothing</u> and keep the title of LEED AP. It is also called LEED AP <u>without</u> specialty, or a legacy LEED AP.

   2) Starting June 2009, a LEED AP can choose to <u>enroll in</u> the new tiered system, accept the GBCI <u>disciplinary policy</u>, and **complete** the prescriptive <u>Credentialing Maintenance Program</u> (**CMP**) for the initial two-year reporting period. After opting in, a LEED AP can use one of the new specialty designations (BD+C, ID+C, O+M) after his name. The LEED AP must opt in before summer of 2011.

   3) Starting June 2009, a LEED AP can choose to <u>opt in</u>, accept the GBCI <u>disciplinary policy</u>, **agree** to CMP, and **pass** Part Two of one of the LEED AP+ specialty exams to become a LEED AP+. The LEED AP only needs to take Part Two of the LEED AP+ exam if he takes the exam <u>by summer, 2011</u>. After opting in, he can use one of the new specialty designations (Building Design and Construction, or BD+C, Interior Design and Construction, or ID+C, Building Operation and Maintenance, or O+M) after his name.

   By choosing Paths 2 or 3 above, a LEED AP will become a LEED AP+ or a LEED AP with Specialty. See detailed information below:
   A LEED AP who passed the <u>old</u> LEED-<u>NC</u> Exam will become a LEED AP <u>BD+C</u>, or a LEED AP <u>BD&C</u>.

   A LEED AP who passed the <u>old</u> LEED-<u>CI</u> Exam will become a LEED AP <u>ID+C</u> or a LEED AP <u>ID&C</u>.

A LEED AP who passed the old LEED-EB Exam will become a LEED AP O+M or a LEED AP O&M.

Credential Maintenance Program (CMP): After completing one of these three choices, a LEED AP will be treated like any other LEED AP+ and will need to pay a $50 fee and take 30 hours of required class every two years to maintain the title of LEED AP+, 6 of the hours must be LEED-specific. The $50 fee is waived for the first two years for a Legacy LEED AP who decides to opt in.

See detailed information on CMP at the following link:
http://www.gbci.org/cmp-wizard

LEED Green Associate, LEED AP+ and LEED Fellow are the three new tiers of professional credentials for LEED professionals. The GBCI started to use these new designations in 2009.

A **LEED Green Associate** is a green building professional with a basic level of LEED knowledge, i.e., a person who has passed the LEED Green Associate Exam and possesses the skills and knowledge to understand and support LEED projects and green building in the areas of design, construction, operation, and maintenance, AND has signed the paperwork to accept the GBCI disciplinary policy.

The LEED Green Associate Exam will NOT test the detailed information for each LEED credit. The test covers core concepts. You need to know the strategies for the overall categories of WE, EA, etc.

However, it IS much easier to understand, digest and organize LEED information, and it will be to your advantage to learn and memorize the LEED information for each specific LEED credit. You will be a much more desirable and useful support member for a LEED project team. You will definitely be able to answer generic LEED questions for the major LEED categories and pass the LEED Green Associate Exam.

You also need to know the codes and regulations related to each of the main LEED categories, i.e., WE or EA, etc., but NOT information for each credit within the category.

Exam Cost: The cost of LEED Green Associate Exam is $200 for USGBC members or full-time students and $250 for non-members per appointment.

Maintenance: A LEED Green Associate will need to pay a $50 fee and take 15 hours of required class every two years to maintain the title, 3 of the hours must be LEED-specific

LEED AP+ is a green building professional with an advanced level of LEED knowledge, i.e., a person who has passed the LEED Green Associate Exam (or Part One of the LEED AP+ exams) and Part Two of a LEED AP+ specialty exam based on one of the LEED rating systems or the equivalent, AND has signed paperwork to accept the GBCI disciplinary policy.

There are five different categories of LEED AP+ specialty exams (Part Two of LEED AP+ exams) and five categories of related LEED AP+ credentials, including:

    LEED AP ID+C (Interior Design and Construction)
    LEED AP O+M (Operation and Maintenance)

LEED AP ND (Neighborhood Development)
LEED AP BD+C (Building Design and Construction)
LEED AP Homes

For example, if you want to become a LEED AP BD+C, you need to pass the LEED Green Associate Exam (Part <u>One</u> of ALL LEED AP+ exams) and the LEED AP BD+C Specialty Exam (Part <u>Two</u> of LEED AP+ exam specializing in LEED BD+C).

Both Part One and Part Two of the LEED AP+ exam have <u>100</u> multiple-choice questions, asking for one, two, three, or even <u>four</u> correct answers (Some questions have <u>five</u> choices).

Exam Cost
1) If you are taking the combined exam (both Part One and Part Two) the cost is <u>$400</u> for USGBC members and <u>$550</u> for non-members.
2) If you are taking only the Part Two specialty exam the cost is <u>$250</u> for USGBC members and <u>$300</u> for non-members.

Maintenance: A LEED AP+ will need to pay a <u>$50</u> fee and take <u>30</u> hours of required class every <u>two</u> years to maintain the title, 6 of the hours must be LEED-specific.

A **LEED Fellow** is a green building professional with an <u>extraordinary</u> level of LEED knowledge, and has made major contributions to green building industry.

Note: The fees listed are the current amounts at the printing of this book. They may be changed by the GBCI at a later point, please check the GBCI website for exact fees.

2. **Why did the GBCI create the three-tier LEED credential system?**
   Answer: The system was implemented to pursue on going improvement and excellence, to assure stakeholders of LEED professionals' current competence and latest knowledge in green building practice, and to meet <u>three</u> prevailing market challenges: <u>staying current</u>, <u>differentiation,</u> and <u>specialization.</u>

3. **Do I need to have LEED project experience to take the LEED exams?**
   Answer: Anyone who is 18 years of age or older can take the LEED Green Associate Exam.

   You also need to agree to the GBCI credential maintenance requirements and the disciplinary policy. See LEED Green Associate Exam candidate handbook for detailed requirements.

   If you are taking the LEED AP+ exam, you must have documented experience on one or more LEED projects within three years of your exam application submittal date. See LEED AP+ exam candidate handbook for detailed requirements. There are online courses that you can take to gain the required LEED experience.

4. **How do I become a LEED AP+? Do I have to take the LEED Green Associate Exam first to become a LEED AP+?**
   Answer: There are <u>four</u> paths to become a LEED AP+. See below. Paths <u>One and Two</u> do <u>NOT</u> require taking the LEED Green Associate Exam first, while Paths Three and Four do.

   Paths One and Two: If you are one of those lucky people who passed one of the three old LEED

AP exams before June 30, 2009, and decided to opt into the new tiered system, you just need to opt in and accept the GBCI disciplinary policy and CMP requirements, and

1) **pass** the Part Two of the LEED AP+ exam before summer of 2011 and **agree** to the CMP

OR

2) **complete** the CMP before summer of 2011.

Path Three: If you have already passed the LEED Green Associate Exam, you only need to pass one of the specialty exams (Part Two of the LEED AP+ exams) to earn the title of one of the LEED AP with specialty.

Path Four: Yes, you can either schedule the two exams separately or back to back in the same sitting. If you fail one of the exams during a single sitting, you only need to reschedule and retake the failed exam part at a different time and for an additional fee.

5. **How many questions do you need to answer correctly to pass the LEED exams?**
   Answer: Many readers have asked me this question before. The short answer is about 60 correct questions or 60% of the 100 total questions. The only official answer you can get is from the GBCI is that they do NOT give out an exact number. So, different people have different opinions. The justification for my answer is outlined below.

   1) The GBCI intends to use 60% of the correct questions as the benchmark for a passing score. Here is a simple calculation:

      LEED exams
      Maximum Score: 200
      Minimum Score: 125
      Difference: 200-125 = 75 points (The score difference between someone who answers everything correctly and someone answers everything incorrectly.)

      75 × 60% = 45 (points)
      125 + 45 = 170 points = passing score

      Total questions: 100
      60% × 100 = 60 = correct answers needed to pass the LEED exams.

   2) However, the exact number may be different depending on the difficulty of the version of the exam that you are taking.

      This is because GBCI wants the LEED passing score to be legally defensible. They may use subject matter experts to set the minimum level of required LEED knowledge and professional psychometricians to analyze the performance of people taking the beta tests, and uses the Angoff Method to decide the final passing score.

      The easier versions of the test will need a higher number of correct answers to pass, and the harder versions of the test will need a lower number of correct answers to pass. So, the correct number of the questions needed to pass the exam is about 60, but probably NOT exactly 60.

If you reach the required level of LEED knowledge, no matter which version of the test you take, you should pass.

6. **What are the key areas that USGBC uses to measure the performance of a building's sustainability?**
   Answer: Integrative Process (IP), Location and Transportation (LT), Sustainable Sites (SS), Water Efficiency (WE), Energy and Atmosphere (EA), Materials and Resources (MR), Indoor Environmental Quality (EQ), Innovation (IN), and Regional Priority (RP).

   Mnemonic: Ian and Larry, Shall We Eventually Make Everything Italic and Red?

7. **How many LEED exams does USGBC have?**
   Answer:
   1) Before 2009, there were three old versions of the LEED exams: LEED New Construction (LEED-NC) v2.2, LEED Commercial Interior (LEED-CI) v2.0, and LEED Existing Building (LEED-EB) v2.0. You just needed to pass one of the three exams to earn the old title of LEED AP.
   2) In 2009, GBCI started to introduce a new three-tier LEED credential system, and related new versions of LEED exams.
      a. The LEED Green Associate Exam is Part One of ALL LEED AP+ exams.

      b. Part Two of LEED AP+ exams includes the following specialties (exam takers just need to choose one):

      LEED AP ID+C (Interior Design and Construction)
      LEED AP Homes
      LEED AP O+M (Operation and Maintenance)
      LEED AP ND (Neighborhood Development)
      LEED AP BD+C (Building Design and Construction)

      Note: GBCI allows LEED AP+'s to have more than one designation. For instance, you can take the LEED AP Homes specialty exam as well as the LEED AP BD+C specialty exam and be considered a LEED AP Homes and a LEED AP BD+C, but you need to take 6 extra LEED Specific CMP hours every two years to maintain the additional specialty designation.

8. **Are LEED exams valid and reliable?**
   Answer: Yes, they are valid because they can measure what they intend to measure. They are reliable because they can accurately measure a candidate's skills.

9. **How many member organizations does the USGBC have?**
   Answer: The USGBC has more than 11,000 member organizations.

10. **How many regional chapters does the USGBC have?**
    Answer: The USGBC currently has 77 chapters.

11. **What is the main purpose of the USGBC?**
    Answer: They are attempting to improve the way a building is designed, built, and used in order to achieve a healthy, profitable, and environmentally responsible building and environment. In turn this is meant to improve quality of life.

12. **What are the guiding principles of the USGBC?**
 Answer: The USGBC emphasizes not only the decisions themselves but also *how* the decisions are made. Their guiding principles are as follows: advocate the **triple bottom lines** (to balance environmental, social, and economic needs or sometimes summarized as planet, people, and profit); build leadership; strive to achieve the balance between humanity and nature; uphold integrity and restore, preserve, and protect the environment, species, and ecosystem; use democratic and interdisciplinary approaches to ensure inclusiveness to achieve a common goal; and openness, honesty, and transparency.

13. **How much energy and resources do US buildings consume?**
 Answer: Per the USGBC and US Department of Energy, buildings consume about 39% of total energy, 74% of electricity and 1/8 of the water in the United States. Buildings also use valuable land that could otherwise provide ecological resources. In 2006, more than 1 billion metric tons of carbon dioxide was generated by the commercial building sector. That is over a 30% increase from the 1990's level.

14. **What is the most important step to get your building certified?**
 Answer: Register your building with USGBC online at usgbc.org.

15. **What are the benefits of green buildings and LEED Certification?**
 Answer: Certification enhances building and company marketability, provides branding opportunities, has positive impact on health and environment, may increase occupant productivity, reduces building operating costs, and helps to create sustainable communities.

16. **Who develops the LEED green building rating systems?**
 Answer: The USGBC committee and volunteers develop the rating systems.

17. **What current reference guides and specific green building rating systems does the USGBC have?**
 Answer: The USGBC has the following reference guides and green rating system portfolios.

 1) *The LEED Reference Guide for Building Design and Construction v4 (BD&C)* covers the following LEED rating systems:
   - LEED BD+C: New Construction
   - LEED BD+C: Core and Shell Development
   - LEED BD+C: Schools
   - LEED BD+C: Retails
   - LEED BD+C: Data Center
   - LEED BD+C: Warehouse and Distribution Center
   - LEED BD+C: Hospitality
   - LEED BD+C: Health Care
   - LEED BD+C: Homes
   - LEED BD+C: Multi-family Midrise

 2) *The LEED Reference Guide for Interior Design and Construction v4 (ID&C)* covers the following LEED rating systems:
   - LEED ID+C: Commercial Interiors

- LEED ID+C: Retails
- LEED ID+C: Hospitality

3) *The LEED Reference Guide for Green Building Operations and Maintenance v4 (O&M)* covers the following LEED rating systems:
   - LEED O+M: Existing Buildings
   - LEED O+M: Retails
   - LEED O+M: Schools
   - LEED O+M: Hospitality
   - LEED O+M: Data Center
   - LEED O+M: Warehouse and Distribution Center

Note: These LEED-EB rating systems are the only systems that cover building operation. All other LEED systems cover building design and construction, but NOT operation.

4) *The LEED for Neighborhood Development Reference Guide v4 (LEED-ND)* covers the following:
   - LEED ND: Plan
   - LEED ND: Built Project

5) **Campus Guide**, 2014 Edition. See the link below:
   http://www.usgbc.org/campusguidance

The complete sample LEED v4 forms are available in the credit library at the following link: http://www.usgbc.org/credits

Refer to this link for various LEED rating systems as well.

After May 31, 2014, ALL LEED exams will be upgraded to LEED v4.
After May 31, 2015, ALL buildings have to be registered under LEED v4.

18. How does LEED fit into the green building market?
Answer: LEED fits into this market by providing voluntary, market-driven, and consensus-based rating systems. Based on accepted environmental and energy principles, LEED maintains a balance between emerging concepts and established practices. Green buildings developed under the LEED rating systems can reduce operating costs and create branding and marketing opportunities for buildings and organizations. They are good for the environment and public health, increase occupant productivity, and create sustainable communities.

19. What are the benefits of LEED certification for your building?
Answer:
1) The building is qualified for various government initiatives.
2) The building can obtain USGBC (third party) validation of achievement.
3) Everyone involved can be recognized for his or her commitment to environmental issues.
4) There are branding opportunities and market exposure through media, Greenbuild conferences, USGBC, cases studies, etc.

20. What is the procedure of LEED certification for your building?
   Answer:
   1) Go to www.gbci.org to <u>register</u> your building. This is the <u>most important</u> step and better if done early on.
   2) Document that your building meets prerequisites and a minimum number of points to be certified.
   3) Refer to the LEED project checklist for the points needed for various levels of LEED certification.

21. How much is the building registration fee and how much is the building LEED certification fee?
   Answer: The building <u>registration</u> fee is $900 for USGBC members and $1,200 for non-members.

   The building <u>certification</u> fee varies, but it starts at $2,250 for USGBC members and $2,750 for non-members.

22. What are LEED's system goals?
   Answer: LEED's system goals or **"Impact Categories"** are as follows:
   - global **climate change**
   - social equity, environmental justice, and **community** quality of life
   - individual **human health** and well-being
   - **greener economy**
   - **biodiversity** and ecosystem
   - **water resources**
   - sustainable and regenerative **material resources** cycles

   These items also answer the question: "What should a LEED project accomplish?"

23. How are LEED credits allocated and weighted?
   Answer: Credits that can contribute to LEED's **"Impact Categories"** are given more points. These impact categories are weighted through a consensus driven process and are as follows:
   - global **climate change** (35%)
   - social equity, environmental justice, and **community** quality of life (5%)
   - individual **human health** and well-being (20%)
   - **greener economy** (5%)
   - **biodiversity** and Ecosystem (10%)
   - **water resources** (15%)
   - sustainable and regenerative **material resources** cycles (10%)

   The USGBC uses three **association factors** to measure and scale credit outcome to a given impact category component.
   1) **Relative efficacy** measures whether a credit outcome has a positive or negative association with a given Impact Category component, and how strong that association is. They're rated as follows:
      - no association
      - low association
      - medium association
      - high association
      - negative association

2) **Benefit duration** measures how long the benefits or consequences of the credit outcome will last:
   - 1-3 Years
   - 4-10 Years
   - 11-30 Years
   - 30+ Years (Building/Community Lifetime)

3) **Controllability of effect** indicates which individual is most directly responsible for achieving the expected credit outcome. The more a credit outcome depends on active human effort, the less likely it will be achieved with certainty, and the credit will have fewer points. Less human effort equals more points.

The USGBC simplifies the weighting process of points into a **scorecard computed as follows**:
- **100 base points** for the base LEED Rating System
- **1 point minimum** for each credit
- **whole points** and no fractions for LEED points

See detailed discussions in the FREE PDF file entitled "LEED v4 Impact Category and Point Allocation Development Process" at the following link:
http://www.usgbc.org/sites/default/files/LEED%20v4%20Impact%20Category%20and%20Point%20Allocation%20Process_Overview_0.pdf

24. Are there LEED certified products?
Answer: ALL LEED systems certify buildings or projects, NOT products. Products can contribute to a LEED project, and sometimes a product's data is required as part of a LEED submittal package.

Chapter 3
Introduction to The LEED Green Associate Exam

1. **What is new for LEED v4?**
   Answer: LEED v4 is a major revision and has two new credit categories: Integrative Process as well as Location and Transportation. Many existing credits have different point values. Some of the criteria for obtaining credits have also changed.

   The numbering system for the credit has been omitted. For example, "SSc5.1: Site Development: Protect or Restore Habitat" is now simplified as "SS: Site Development: Protect or Restore Habitat." SS stands for Sustainable Sites development. The USGBC has also changed the IEQ category to EQ category.

   In addition to the rating system for New Construction, Core and Shell, and Schools, LEED BD+C reference guide have added information for the following rating systems: Retail, Data Center, Warehouse and Distribution Center, Hospitality, and Healthcare.

   A LEED AP without specialties (Legacy LEED AP) can no longer earn a point for the Innovation Credit. Only a LEED AP with a specialty appropriate to the project who works as a principal participant of the project team can earn one point for the Innovation Credit.

2. **What is the scope of the LEED Green Associate Exam?**
   Answer: Per the USGBC, the content of the LEED Green Associate Exam is limited to understanding the following items.

a. **fundamental credit intents, strategies, requirements, technologies, and submittals for the eight major credit categories** (These are Integrative Process (IP), Location and Transportation (LT), Sustainable Sites (SS), Water Efficiency (WE), Energy and Atmosphere (EA), Materials and Resources (MR), Indoor Environmental Quality (EQ), Innovation (IN), and Regional Priority (RP), as well as innovation in upgrades, maintenance, and operations.)

b. **process of LEED application and opportunities for synergy**
   Synergy is the combined effects of two or more agents or forces greater than the sum of their individual.
   The LEED Green Associate Exam may test the following aspects:
   - site, budget, schedule, program, and other requirements
   - hard, soft, and life-cycle costs
   - environmental building news, **USGBC**, **NRDC** (Natural Resources Defense Council), and other green resources
   - **Green Seal**, **SMACNA** (Sheet Metal and Air Conditioning Contractors National Association Guidelines), **ASHRAE** (American Society of Heating, Refrigeration, and Air-conditioning Engineers), and other standards for LEED credit
   - waste management, environmental quality, energy, and other interaction between LEED

credits
- **CIR** (Credit Interpretation Requests/Rulings) and previous samples leading to extra points
- project registration and **LEED Online**
- score cards for LEED
- supplementary documentation, project calculations, and other letter templates
- LEED credit **strategies**
- property, project, and LEED boundaries
- minimum LEED certification program requirements and/or **prerequisites**
- certification goal and preliminary rating
- opportunities for the same building to get multiple certifications, i.e., commercial interior as well as core and shell
- certified building in LEED neighborhood development
- operations and maintenance for certified new building construction
- requirements for occupancy (e.g., an existing building must be fully occupied for 12 continuous months as described in minimum program requirements)
- logo usage, trademark usage, and other USGBC policies
- requirements for receiving LEED AP credit

c. **Site factors of a project**
   1) **Connectivity** for Communities
      a) Carts, shuttles, car-sharing membership (e.g. Zipcar™), bike storage, public transportation, fuel efficient vehicle parking, parking capacity, carpool parking, and other means to improve **transportation**
      b) Ramps, crosswalks, trails, and other means to improve **pedestrian access** and circulation

   2) **Zoning requirements** (calculations of site area, floor to area ratio, and other density components, building footprint, open space, development footprint, construction limits, and specific landscaping restrictions)

   3) **Development**
      Heat Islands including albedo, emissivity, non-roof/roof heat island effect; green roofs, Solar Reflectance Index (SRI), etc.

d. **Management of water**
   1) Understand quality and types of water, i.e., blackwater, graywater, storm water, and potable water

   2) Understand water management, i.e., use low-flush fixtures such as water closets, urinals, sinks, lavatory faucets, and showers to reduce water use; calculations of FTE (Full Time Equivalent); baseline water demand; irrigation; rainwater harvesting

e. **Project system and related energy impact**
   1) Ozone depletion, chlorofluorocarbon (CFC) reduction, no refrigerant option, fire suppressions without halons or CFC's, phase-out plan, Hydrochlorofluorocarbons (HCFC), and other Environmental Concerns
   2) Green-e providers, off-site generated, renewable energy certificates, and other Green Power

f. **Project materials acquisition, installation and management**

1) Commingled, pre-consumer, post-consumer, collection, and other requirements regarding recycled materials
2) Locally/regionally manufactured and harvested materials
3) Accounted by weight or volume, written plan, polychlorinated biphenyl (PCB) removal and asbestos-containing materials (ACM) management, reduction strategies, and other requirements regarding construction waste management

g. **Project team coordination, stakeholder involvement in innovation and regional design**
   1) Civil engineer, landscape architect, architect, heating-ventilation-air-conditioning (HVAC) engineer, contractor, facility manager, and related integrated project team criteria
   2) Building reuse, material lifecycle, and other durability, planning, and management
   3) Appropriate and established requirements, regional green design and construction, and other innovative and regional design measures

h. **Codes and regulations, public outreach and project surroundings:** building, electrical, mechanical, plumbing, fire protection codes, etc.

i. **Ability to support the coordination of team and project and assist** with the process of gathering the necessary requirements and information for the **LEED process** and coordinating the different job functions for **LEED certification**

j. **Ability to support the process of LEED implementation**

k. **Ability to support technical analyses for LEED credits**

3. What is the latest version of LEED and when was it published?
   a. **When was LEED v1.0 released?**
   Answer: This version was first launched in August 1998, but officially released in 1999.

   b. **When was LEED-NC v2.0 released?**
   Answer: This version was first published in 1999, but officially released in March 2000.

   c. **When was LEED-NC v2.1 released?**
   Answer: This version was officially released in November 2002.

   d. **When was LEED-NC v2.2 released?**
   Answer: This version was first published in 2003, but officially released in 2005.

   e. **When was LEED v3.0 released?**
   Answer: In April 2009, the USGBC launched **LEED v3.0**, which included a new building certification model, a new LEED Online, and an improvement to the LEED rating systems (LEED 2009). LEED v3.0 was a part of the continuous evolution of the LEED building rating system. For LEED v3.0, the USGBC was trying to synchronize the prerequisites and credits across each version of the LEED system, and to create a predictable LEED development cycle (similar to other building codes which are usually updated every three years), a transparent environmental/human impact credit weighting (redistributing the available points in LEED), as well as regionalization (regional bonus credits). Refer to the appendixes of this book for links and more information on LEED v3.0.

f. **When was LEED v4 released?**
Answer: The USGBC released LEED v4 at the GreenBuild International Conference and Expo in November 2013.

4. **How many possible points does LEED v4 have?**
Answer: Starting with LEED v3.0 or LEED 2009, **ALL** updated rating systems have or will have **110** possible points, including **100** possible base points and **10** bonus points. LEED v4 maintains the same number of total points, but changes how many points a credit can get.

The latest scorecards for various LEED rating system can be found at the following link: http://www.usgbc.org/credits

For example, LEED v4 **BD+C: New Construction** has up to 1 point for Integrative Process (IP), 16 points for Location and Transportation (LT), 10 points for Sustainable Sites (SS), 11 points for Water Efficiency (WE), 33 points for Energy and Atmosphere (EA), 13 points for Materials and Resources (MR), 16 points for Indoor Environmental Quality (EQ), 6 bonus points for Innovation (IN), and 4 bonus points for Regional Priority (RP).

LEED v4 **BD+C: Core and Shell** has up to is 1 point for Integrative Process (IP), 20 points for Location and Transportation (LT), 11 points for Sustainable Sites (SS), 11 points for Water Efficiency (WE), 33 points for Energy and Atmosphere (EA), 14 points for Materials and Resources (MR), 10 points for Indoor Environmental Quality (EQ), 6 bonus points for Innovation (IN), and 4 bonus points for Regional Priority (RP).

LEED v4 **BD+C: Schools** has up to 1 point for Integrative Process (IP), 15 points for Location and Transportation (LT), 12 points for Sustainable Sites (SS), 12 points for Water Efficiency (WE), 31 points for Energy and Atmosphere (EA), 13 points for Materials and Resources (MR), 16 points for Indoor Environmental Quality (EQ), 6 bonus points for Innovation (IN), and 4 bonus points for Regional Priority (RP).

LEED v4 **ID+C: Commercial Interiors (CI)** has up to 2 points for Integrative Process (IP), 18 points for Location and Transportation (LT), 12 points for Water Efficiency (WE), 38 points for Energy and Atmosphere (EA), 13 points for Materials and Resources (MR), 17 points for Indoor Environmental Quality (EQ), 6 bonus points for Innovation (IN), and 4 bonus points for Regional Priority (RP).

LEED v4 **for Existing Building Operation and Maintenance** has up to 15 points for Location and Transportation (LT), 10 points for Sustainable Sites (SS), 12 points for Water Efficiency (WE), 38 points for Energy and Atmosphere (EA), 8 points for Materials and Resources (MR), 17 points for Indoor Environmental Quality (EQ), 6 bonus points for Innovation (IN), and 4 bonus points for Regional Priority (RP).

5. **How many different levels of building certification does USGBC have?**
   Answer: The USGBC has four levels of building certification. The four levels and the points required for each are as follows:

Certified	**40**–49	points
Silver	**50**–59	points
Gold	**60**–79	points
Platinum	**80**	points and above

   The level achieved by a building is based on the number of points earned under the LEED green building rating system. The points are calculated using the scorecards or checklists provided by the USGBC.

   No matter which **LEED v4.0** rating system you choose, each LEED v4.0 rating system has **100** base points; a maximum of **6** possible extra points for Innovation (IN), and a maximum of **4** possible extra points for Regional Priority (RP) for each project. There are 6 possible RP points, but you can only pick and choose a maximum of 4 points for each project.

   Refer to the following link for the **Regional Priority Credits (RPC)** for all 50 States. With this spreadsheet, you can locate the RPC for your area by zip code:
   http://www.usgbc.org/rpc

   A LEED project team does not have to do anything special, since LEED v4 Online will automatically decide which RPC your project will get once you enter the project zip code and other information. If you have more than 4 RPCs, then you need to decide which 4 RPCs you want to use for your project.

   The USGBC has been working on developing similar RPC incentives for international projects. RPC is available for many countries

6. **What is the process or basic steps for LEED certification?**
   Answer: The USGBC suggests ten basic steps for LEED certification:
   1) **Initiate the discovery phase** by gathering information, doing research and analysis, and holding a goal-setting workshop.
   2) **Select a LEED rating system.** See *Rating System Selection Guidance*, as well as *Further Explanation* under each credit for reference.
   3) **Check minimum program requirements (MPRs)** for the selected rating system. This information is available in the related reference guide and on the USGBC website.
   4) **Establish project goals** by holding a goal-setting workshop (see Integrative Process Credit) for the project team members and the owner, which should include representatives from the design and construction disciplines.
   5) **Define the LEED project scope.** Consider shared facilities or off-site or campus amenities that may be used by project occupants, map the LEED project boundary, and investigate special certification programs such as the **Volume Program** or the **Campus Program**.
   6) **Develop the LEED scorecard** and focus on those credits with the highest value in the long term. Also seek synergistic benefits, set the target LEED certification level (Certified, Silver, Gold, or Platinum), be sure to meet **prerequisites,** and include several extra points as a buffer.
   7) **Continue the discovery phase** by completing additional research and analysis. Reassemble the team and owner occasionally to coordinate.

8) **Continue the iterative process** with research and analysis as well as workshops until the solution satisfies both the team and the owner.
9) **Assign roles and responsibilities.** Select a leader for the LEED application and documentation process and assign primary and supporting roles to appropriate team members for each credit. Establish regular meeting dates and open clear communication channels.
10) **Perform quality assurance reviews and submit for certification.** This should be a thorough quality control check making sure numeric values (e.g., site area) are consistent across credits.

The USGBC publishes three guides for LEED v4 certification. You should read these guides at least *three* times. They will not only help you pass a LEED exam on the first try, but also benefit your real LEED project.

- *Guide to LEED v4 Certification: Commercial*

   Commercial projects include four main steps:
   a. **Register** the project by completing key forms and submitting payment.
   b. **Apply** for LEED certification by submitting an application via LEED Online and paying for a certification review fee.
   c. **Review** takes place by GBCI.
   d. **Certification** is either "denied" or "achieved."

   See the following link:
   http://www.usgbc.org/cert-guide/commercial

   Note: Registration is an important step. The project administrator has access to the CIR database and LEED Online after registration.

   As part of the registration, the LEED project team will *have to* agree to report *post-occupancy water and energy use*. This is to allow the USGBC to have a better understanding of the relationship between building performance and LEED credits. There are several ways to achieve the reporting requirements, including signing up for LEED O&M, OR signing a waiver to allow the USGBC to acquire the data applicable directly from the utility company.

- *Guide to LEED v4 Certification: Homes*

   See the following link:
   http://www.usgbc.org/cert-guide/homes

   Residential projects include four main steps (Please note the differences from commercial projects):
   a. **Register** the project by *selecting your team*, completing key forms, and submitting payment.
   b. **Verify** your project milestones and achievements via the on-site verification process. A mid-construction verification site visit, sometimes called the "pre-drywall" visit by the **Green Rater** and **Energy Rater** is mandatory. A final construction verification visit is required after construction and landscape is complete. During this visit the **energy rater** also conducts the required performance testing.
   c. **Review** takes place in stages. First, submit the necessary information, calculations and documentation to your **Green Rater**. The Green Rater then submits the appropriate

documentation to the **LEED for Homes Provider** for their *quality assurance* review. The LEED for Homes Provider finishes their review and then submits the documentation to GBCI for *certification* review.
   d. **Certification** is either "denied" or "achieved."

For LEED for Homes, you need to deal with a LEED for Homes Provider, a Green Rater, and an Energy Rater.

A LEED for Homes Provider is the referee when it comes to who is able to be a **Green Rater** for a LEED for Homes project. The Provider is responsible for hiring, training, and overseeing the Green Raters. The USGBC requires that each Provider have a quality assurance protocol for its Green Raters. Homebuilders who intend to achieve LEED for Homes certification **must contact a LEED for Homes Provider** organization.
Refer to the following link for a complete list of LEED for Homes Provider organizations in the United States:
http://www.usgbc.org/organizations/members/homes-providers

A Green Rater is a professional who has shown qualifications for energy rating certification, and demonstrates the ability to deal with a home's energy systems and performance. Certification can be obtained from accreditation bodies such as ENERGY STAR, RES NET, BPI, etc. Green Raters provide the required *on-site* verification.

An Energy Rater initiates a *performance* test which is mandatory for the LEED for Homes rating system. The largest body of energy raters is called **Home Energy Raters (HERS Raters).** The **Residential Energy Services Network (RES NET)** administers HERS Raters credentials.

- *Guide to LEED v4 Certification: Volume supplement*

   See the following link:
   http://www.usgbc.org/cert-guide/volume

The LEED Volume Program is available for the following LEED rating systems:
- LEED for New Construction (LEED NC)
- LEED for Commercial Interiors (LEED CI)
- LEED for Retail: New Construction (LEED RN)
- LEED for Retail: Commercial Interiors (LEED RI)
- LEED for Existing Buildings: Operations and Maintenance (LEED EB: O+M)
- projects undergoing recertification

Participants in the **LEED Volume Program** finish precertification of a prototype, and then the projects based on this prototype can easily earn a common set of credits. This can save a lot of time and money for an organization.

All Volume Program participants are organizations that own, lease or manage real estate. Architects, consultants, and contractors are not eligible.

There are two forms of CIRs under the Volume Program

- A **prototype CIR** is applicable to all buildings in the participant's portfolio for the prototype.
- A **volume project CIR** is applicable only to the specific volume project for which it was submitted.

See detailed discussion on CIRs on the following pages.

See the following links for more information on LEED certification and certification fees:
http://www.usgbc.org/cert-guide/fees
http://www.usgbc.org/cert-guide
http://www.usgbc.org/leed/certification

USGBC offers the following resources and tools to assist LEED certification:
- **Credit Library**
  http://www.usgbc.org/credits
- **Addenda database**
  http://www.usgbc.org/leed-interpretations
- **Pilot Credit library**
  http://www.usgbc.org/pilotcredits
- **Regional Priority Credit lookup**
  http://www.usgbc.org/rpc

7. **What does the registration form include?**
   Answer: The registration form includes account login information, project contact information, project details (type, title, address, owner, gross square footage, budget, site condition, current project phase, scope, occupancy, etc.), and the confidentiality status of the project.

8. **What is precertification?**
   Answer: Precertification is mainly for the Core and Shell program. It is a formal recognition of a project where the owner intends to seek Core and Shell certification, and gives the owner/developer a marketing tool to attract financers and tenants. A project needs to meet a scorecard requirement for precertification. Precertification is granted after an early design review by the USGBC, and the review usually takes less than one month. Precertification costs $2,500 for USGBC members and $3,500 for non-members. This amount does NOT cover the fee for certification, and does NOT guarantee the final certification of the project.

9. **What is a CIR?**
   Answer: CIR stands for **Credit Interpretation Request** or **Credit Interpretation Ruling**, depending upon the context. Most of the time CIRs mean Credit Interpretation Rulings. CIRs are *not* precedent setting.

   LEED Interpretation is similar to CIRs, but these *are* precedent setting.

10. **When do you submit a CIR?**
    Answer: When there are conflicts between two credit categories or the USGBC Reference Guide does *not* give you enough information a CIR can be submitted. The building registration fee covers two free CIRs, but after 11/15/05, it costs $200 to submit each CIR.

11. What are the steps for submitting a CIR?
Answer: The steps and some pointers for submission are as follows:
1) Check your project against each LEED credit or prerequisite.
2) Check the USGBC Reference Guide.
3) Review the GBCI and USGBC websites for previous CIRs and LEED Interpretations.
4) If you cannot find a similar CIR, then submit a new CIR to GBCI online.
5) Do *not* mention the contact information, name of the credit, or confidential information.
6) Submit only the essential and required information, and do not submit it in a letter format.
7) Submit one CIR for each prerequisite or credit.
8) Do not include any attachments.
9) Include details and background information, limited to 600 words (4,000 max characters including spaces).

12. Will a CIR guarantee a credit?
Answer: No, a CIR only provides feedback, and it will *not* guarantee a credit. Also, no credit will be awarded in the CIR process.

13. Which tasks are handled by the GBCI and what tasks are handled by the USGBC?
Answer: In 2008, the Green Building Certification Institute (GBCI) spun off from the USGBC.

The USGBC still handles the online tools, LEED rating system development, and related educational offerings, etc.

The GBCI took over some responsibilities from the USGBC and handles building LEED certification and LEED professional accreditation.

For LEED building certification, the GBCI oversees 10 organizations including Lloyd's Register Quality Assurance and Underwriters Laboratories, which manage the project review process.

This separation of the tasks for the USGBC and the GBCI meets the protocols of the American National Standard Institute (ANSI) and International Organization for Standardization (ISO), and makes the building certification a true third-party process.

14. What are MPRs?
Answer: MPRs are **Minimum Program Requirements**. A project must meet MPRs to qualify for LEED certification.

MPRs serve three goals:
1) They provide clear guidance for customers.
2) They maintain LEED program integrity.
3) They make the LEED certification process easier.

MPRs include some very basic requirements, including the following:
1) A LEED project must be must be in a permanent location on existing land.
2) A LEED project must have a *reasonable* site boundary, and the building **gross floor area** to **gross land area** within the LEED project boundary must be 2% or higher.
3) The building must comply with project size requirements. For example, the building must be 1,000 s.f. minimum for LEED BD+C and O+M rating systems, and 250 s.f. minimum for the LEED ID+C rating system. For LEED Neighborhood Development rating systems, the LEED

project should be no larger than 1,500 acres and contain a minimum of two habitable buildings.

Refer to the link below for detailed information:
http://www.usgbc.org/credits/homes/v4/minimum-program-requirements

15. Which LEED rating system should I choose?

Answer: The USGBC has published a *LEED v4 User Guide* to assist you in choosing the right rating system for your project.

The **LEED BD+C** rating systems should be used for new construction or a major renovation with at least 60% of the project's gross floor area that must be complete by the time of certification (except for LEED BD+C: Core and Shell).

A **major renovation** means major building envelope changes, significant HVAC renovation, and interior rehabilitation.

LEED for Interior Design and Construction should be used for interior spaces that are a *complete interior fit-out* with at least 60% of the project's gross floor area that must be complete by the time of certification.

LEED for Building Operations and Maintenance should be used for existing buildings undergoing improvement work or little to no construction.

LEED for Neighborhood Development should be used for new land development projects or redevelopment projects containing residential uses, nonresidential uses, or a mix. The USGBC recommends that a minimum of 50% of total building floor area should be new construction or major renovation.

Each of the aforementioned main rating systems includes more detailed and specific rating systems, such as LEED BD+C: New Construction and Major Renovations, LEED BD+C: Core and Shell Development, and LEED BD+C: Schools, etc.

The USGBC uses a **40/60 rule** to assist you when several rating systems seem to be acceptable for a project:

- Do not use a rating system if it is appropriate for less than 40% of the gross floor area of a LEED project building or space.
- Use a rating system if it is appropriate for more than 60% of the gross floor area of a LEED project building or space.
- A project team can decide which system to use if an appropriate rating system falls between 40% and 60% of the gross floor area.

See *LEED v4 User Guide* at the following link:
http://www.usgbc.org/sites/default/files/LEED%20v4%20User%20Guide_Final_0.pdf

Chapter 4
LEED Green Associate Exam
Overall Technical Review

In this chapter, we introduce the LEED certification process, key components of the LEED rating system, as wells as the purpose, core concepts, strategies, incentives, recognitions, and regulations for each LEED credit category.

1. **What do green buildings address?**
   Green buildings mitigate <u>d</u>egradation of ecosystems/habitat and <u>r</u>esource depletion, reduce <u>c</u>osts of operating and owning living and work spaces, improve <u>o</u>ccupant productivity and comfort, <u>in</u>door environmental quality, and reduce water <u>c</u>onsumption.
   Mnemonic: DR. COIN C. (See underlined letters in the sentence above.)

   Note: Some people like to use mnemonics and think they are very helpful, while others don't. We provide them as an option for you. If you do not like to use mnemonics, just ignore them and read the information a few times to become familiar with the material.

2. **Key stake holders and an integrative approach**
   Green buildings employ an **integrative approach** and encourage the participation of **key stakeholders**, including the following:

   1) **The client** consists of the facilities management staff, facilities O&M staff, community members, owner, and planning staff.

   2) **The design team** includes the civil engineer, landscape architect, architect, mechanical and plumbing engineer, electrical engineer, structural engineer, commissioning authority, and energy and daylighting modeler.

   3) **The builders** are general contractors, EMP subcontractors, cost estimators, construction managers, and product manufacturers.

   There are many **benefits** to an **integrative approach**, including <u>b</u>etter indoor air quality, <u>i</u>mproved occupant performance, <u>r</u>educed operating and maintenance costs, <u>d</u>urable facilities, reduced environmental <u>i</u>mpact, <u>p</u>otentially no increase in construction cost, <u>o</u>ptimized return on investment, and <u>o</u>pportunity of <u>l</u>earning.

3. **The mission of the USGBC**
   The **mission** of the USGBC is
   1) to improve the quality of life;
   2) to change the way we design, build, and operate communities and buildings; and
   3) to create a healthy, prosperous, socially, and environmentally responsible environment.

4. **The structure of LEED rating system**

   In addition to the main credit categories of IP, LT, SS, WE, EA, MR, EQ, IN, and RP, the following factors are very important to the LEED rating systems: awareness and education, green buildings and infrastructure, and neighborhood design and pattern.

   The structure of the LEED rating system includes three tiers:
   1) The main credit categories are IP, LT, SS, WE, EA, MR, EQ, IN, and RP.
   2) Each main credit category includes prerequisites and credits.
   3) Each prerequisite or credit includes intents and requirements (or paths to achieve LEED points).

5. **LEED certification tools**

   LEED certification tools include the following:
   1)   USGBC Reference Guides
   2)   LEED Rating systems
   3)   LEED Online
   4)   LEED Scorecard
   5)   LEED letter template
   6)   CIRs (Credit Interpretation Rulings)
   7)   LEED Case Studies
   8)   USGBC and GBCI websites (www.USGBC.org and www.GBCI.org)

   See discussions on the LEED certification process in chapter 3 also.

   The GBCI requires that the project team submit an overall narrative and complete the LEED Online documentation for the LEED certification application. The general documentation includes project timeline and scope, site conditions, project team identification, usage data and occupant, etc.

   The project's overall narrative includes the team, building, site, and applicant's organization.

   LEED Online includes the project's detailed information and template/complete documentation requirements for completing prerequisites and credits. These include my action items, potential LEED ratings, attempted credit summary (not awarded, anticipated, denied, and total attempted), appealed credit summary, and credit scorecard, etc. It also includes embed tables and calculators to assure accuracy and completeness of the submittal package.

   LEED Online also includes definition for a **declarant**, a Licensed Professional Exemption Form, and related information. A **declarant** is the team member(s) who signs off on the documents and indicates who is responsible for each credit or prerequisite.

   A licensed professional exemption form is the form for a team member who is a licensed/registered landscape architect, architect, or engineer to use as a tool to request waiver for eligible submittal requirements. Licensed professional exemptions are shown in the related LEED credit section of LEED Online.

   With LEED Online, project teams can upload support files, submit applications for reviews and CIRs, receive feedback from the reviewer, contact customer service, generate specific reports for the project, and obtain project LEED certification, as well as gain access to additional LEED resources like tutorials, FAQs, sample documentation, and offline calculators.

If you have multiple projects, you can access all of them via LEED Online.

The GBCI also issues LEED certificates for successful projects via LEED Online.

The credit scorecard includes the project name, address, and a list of all points in various credit categories. With LEED Online, you can choose to collapse all credit categories or view a printer friendly scorecard.

The LEED letter template includes your name, your company name, specific and credit-related project information like project site area (s.f.), gross building area (s.f.), and the credit path you choose for each credit.

As a LEED professional, you also need to know the **synergies** between the main credit categories of IP, LT, SS, WE, EA, MR, EQ, IN, and RP. Almost every LEED exam will have a large percentage of test questions related to **synergies.**

6. Specific technical information

The USGBC workshops may not give you enough information to pass the LEED Green Associate Exam, so I have included part of the specific LEED v4 BD&C technical information to fill in these blanks. This is only a partial version (about 70%) of the information based on LEED 2009 BD&C. Most of the specific information that follows is common or useful to ALL LEED systems, and I think the information will help you.

Based on my research and the USGBC workshops that I have attended, the LEED Green Associate Exam, will unlikely require you to differentiate each detailed credit, and how many exact points a credit can earn. However, the GBCI will want you to know the information contained within the main category of SS for example. Thus, you DO need to know some technical information for the exam. You also definitely need to know how many points your project will need to become LEED Certified, LEED-Silver, LEED-Gold and LEED-Platinum, and the total possible number of IN points, RP points, etc.

There are a few detailed credits for one or two rating systems only. I do not include these because the LEED Green Associate Exam is unlikely to test this kind of specialized information.

Some credits are for **several rating systems** and they do have some generic or common LEED technical information with some minor variations between **them**. The GBCI may test these, so I have included them in the book.

I still present the information in the credit-by-credit format, because it helps to organize the information per the USGBC rating system in a useful way and makes it easier to understand. If I presented all the information for each main category altogether without breaking it down, it would be confusing and difficult for you to understand and digest. Knowing more of what the exam will test is better than knowing less.

Chapter 5
Integrative Process (IP)

*IP Prerequisite (IPP): **Integrative Project Planning and Design**

Mandatory for
Healthcare (0 points)

Note:
- *We add an asterisk (*) for special information that is not common to ALL LEED BD+C systems.*

- *Detailed discussions have been omitted since this credit is for Healthcare only, and is unlikely to be tested on the LEED Green Associate Exam.*

- *Since the LEED Green Associate Exam is unlikely to test you on very specific, technical, or detailed information, we focus on the generic LEED and green building concept information. For specific, technical, or detailed information on how to achieve each prerequisite and credit, see my other book, LEED BD&C Exam Guide.*

IP Credit (IPC): **Integrative Process**

Applicable to
- New Construction (1 point)
- Core and Shell (1 point)
- Schools (1 point)
- Retail (1 point)
- Data Centers (1 point)
- Warehouses and Distribution Centers (1 point)
- Hospitality (1–5 points)
- Healthcare (1–5 points)

Purpose

Uses an early analysis of the interrelationships among systems to support cost-effective, high-performance project outcomes.

Credit Concept

An integrative process (IP) is a comprehensive approach. It focuses on looking for synergies among components and systems, as well as the mutual advantages to achieve high levels of human comfort, building performance, and environmental benefits. The process involves coordination, rigorous questioning, challenging typical project assumptions, and collaboration to enhance the effectiveness and efficiency of all systems.

This credit encourages early integration and goes beyond checklists. It will be most effective in improving performance at an early stage when clarifying performance goals, the owner's aspirations, and project needs.

An integrative process includes three phases.
- **Discovery** is the most important stage and an expansion of pre-design. It is essential for achieving cost-effectiveness and a project's environmental goals.
- **Design and construction** begins with schematic design. Unlike a conventional project, IP design will use system interactions that were found during discovery and incorporate them.
- **The period of occupancy, operations, and performance feedback** sets up feedback mechanisms and measures performance. Feedback is crucial to determining success in achieving performance targets, informing building operations, and taking related corrective actions.

This credit is an introduction to the comprehensive integrative process and the interactions among all building and site systems. It rewards project teams for applying an integrative approach to water and energy systems. By studying building system interrelationships, project teams can ideally discover unique opportunities for increased building performance, innovative design, and greater environmental benefits, as well as earn more LEED points. By understanding synergies, teams will optimize resources and save time and money in both the short and the long term. Finally, the integrative process can help avoid design changes during the construction documents phase and the related delays, and costs as well as reduce change orders during construction.

Through the integrative process, project teams can identify interrelated issues and develop synergistic strategies, and more effectively use LEED as a comprehensive tool. When applied properly, the integrative process reveals the synergies, instead of individual items on a checklist.

Synergies
- LT Credit (LTC): Quality Transit
- LTC: Reduced Parking Footprint
- SS Credit (SSC): Site Assessment
- SSC: Open Space
- SSC: Rainwater Management
- SSC: Heat Island Reduction
- SSC: Light Pollution Reduction
- WE Prerequisite and Credit (WEP and WEC): Outdoor Water Use Reduction
- WEP and WEC: Indoor Water Use Reduction
- WEC: Cooling Tower Water Use
- EA Prerequisite (EAP): Fundamental Commissioning and Verification
- EAP: Minimum Energy Performance
- EA Credit (EAC): Optimize Energy Performance
- EQ Prerequisite (EQP): Minimum Indoor Air Quality Performance
- EQ Credit (EQC): Enhanced Indoor Air Quality Strategies
- EQC: Thermal Comfort
- EQC: Daylight
- EQC: Quality Views

Chapter 6
Location and Transportation (LT)

Overall Purpose

The LT category is an offshoot of the Sustainable Sites (SS) category. The current SS category focuses on on-site ecosystem services, while the LT category emphasizes the existing features of the surrounding community and their impact on occupant behavior and environmental performance.

Well-located buildings utilize existing utilities and infrastructure, reduce ecological and material costs, and encourage alternatives to private automobile use, such as biking, walking, public transit, and vehicle shares. LT strategies can enhance health by encouraging daily physical activity and reduce greenhouse gas emissions from transportation.

Consistent Documentation

Walking and bicycling distances are measurements of distance from a point of origin to a destination, such as the nearest train station. They must be measured along a *safe* path, NOT a straight-line radius from the origin that does not follow the path.

Safe walking paths include sidewalks, crosswalks, all-weather-surface footpaths, or equivalent pedestrian facilities.

Safe bicycling paths include streets with low target vehicle speed, on-street bicycle lanes, and off-street bicycle paths or trails.

Total vehicle parking capacity is the number of all off-street spaces that the project building's users can use, both inside and *outside* the project boundary.
The following parking spaces should *not* be included:
- bicycle or motorbike spaces
- on-street (parallel or pull-in) parking spaces
- parking spaces for inventory and fleet vehicles, unless these vehicles are regularly used by employees for commuting as well as business purposes

Preferred parking spaces are closest to the main entrance of the project, exclusive of handicap spaces. For a multiple-level parking facility, locate preferred spaces on the level closest to the main entrance of the building.

LT Credit (LTC): **LEED for Neighborhood Development Location**

Applicable to
New Construction (8-16 points)
Core and Shell (8-20 points)
Schools (8-15 points)
Retail (8-16 points)
Data Centers (8-16 points)
Warehouses and Distribution Centers (8-16 points)
Hospitality (8-16 points)
Healthcare (5-9 points)

Purpose
Uses an early analysis of the interrelationships among systems to support cost-effective, high-performance project outcomes.

Credit Concept
This credit allows the project to earn some points for locating the building in a certified LEED for Neighborhood Development (LEED ND).

The LEED ND rating system integrates principles of new urbanism, smart growth, green building design, and construction to promote healthy, sustainable, and equitable places for neighborhood workers, residents, and visitors. Certified neighborhoods have to exhibit many sustainability features, such as transit access, walkability, connectivity, sensitive land protection, and shared infrastructure.

By selecting a project site in a LEED ND certified neighborhood or plan, project teams have demonstrated a commitment to the goals of the LT credit category: linkages with the surrounding community and excellent building location. The criteria for this credit is that the project be LEED ND certified (not just registered), to meet the goals of the LT category. This credit is an alternative to the pursuit of the individual LT credits.

Synergies
- Projects applying for this credit are not eligible to earn points under other LT credits.

LTC: Sensitive Land Protection

Applicable to
New Construction (1 point)
Core and Shell (2 points)
Schools (1 point)
Retail (1 point)
Data Centers (1 point)
Warehouses and Distribution Centers (1 point)
Hospitality (1 point)
Healthcare (1 point)

Purpose
Reduces a building's environmental impact and avoids the development of environmentally sensitive land.

Credit Concept
Sensitive land provides many environmental and human health benefits. Agricultural areas assist with rainwater management and produce food; floodplains support diverse fauna and flora, provide flood protection, and supply rich agricultural soil; endangered or imperiled species habitat supports biodiversity; and water bodies and wetlands can buffer against sequester carbon, flooding, and manage rainwater runoff.

We need to effectively manage the ecosystem benefits and services provided by these sensitive land types. Development of this land places people and property in danger, and harms the ecology of the area. For instance, agricultural land conversion limits local opportunities to produce food, and placing a building close to a marsh or in a floodplain increases its risk of damage from rising sea level or floods.

Selecting a site that has previously been developed and limiting the building's footprint to the previously developed area is one strategy to lessen the environmental consequences of a building. It also encourages the investment in existing neighborhoods and reuse of existing built infrastructure.

If it is not possible to develop entirely on previously developed land, project teams can choose not to disturb sensitive land types to achieve compliance.

Synergies
- LTC: High-Priority Site
- LTC: Surrounding Density and Diverse Uses
- LTC: Reduced Parking Footprint
- SSC: Site Assessment
- SSC: Rainwater Management
- SSC: Site Development—Protect or Restore Habitat

LTC: High-Priority Site

Applicable to
New Construction (1-2 points)
Core and Shell (3 points)
Schools (1-2 points)
Retail (1-2 points)
Data Centers (1-2 points)
Warehouses and Distribution Centers (1-2 points)
Hospitality (1-2 points)
Healthcare (1-2 points)

Purpose
Encourages locating projects in places with development constraints and supports the health of the adjacent areas.

Credit Concept
Many governments and communities have identified high-priority redevelopment sites. Instead of development in greenfields and environmentally sensitive areas, reusing these sites has many environmental advantages. It can bring economic and social benefits to the surrounding community and revitalize the neighborhood. Reusing existing infrastructure can also achieve savings.

There are three paths to achieve this credit:
1) **Encourage investment in historic areas.** This is a proven strategy for enhancing or maintaining community character. Adaptive reuse can also reduce urban sprawl.

2) **Reward the location of certain appropriate projects on economically disadvantaged or depressed sites.** Perceived economic barriers or stigmas can cause vacant or underutilized sites in many low-income communities. This path promotes the economic and social revitalization of neighborhoods.

3) **Promote the redevelopment of contaminated sites.** Removing hazardous materials from a site's groundwater, soil, and any existing buildings, can improve environmental health and reduce the exposure of wildlife and humans to environmental pollution. It often reduces the footprint of the project's elements since a redevelopment site uses an average of 78% less land than the same or similar project built on a greenfield.

Synergies
- MRC: Building Life-Cycle Impact Reduction
- SSP: Environmental Site Assessment
- LT credit category (all credits)

LTC: Surrounding Density and Diverse Uses

Applicable to
    New Construction (1-5 points)
    Core and Shell (1-6 points)
    Schools (1-5 points)
    Retail (1-5 points)
    Data Centers (1-5 points)
    Warehouses and Distribution Centers (1-5 points)
    Hospitality (1-5 points)
    Healthcare (1 point)

Purpose
    Conserves land and protects wildlife habitat and farmland by encouraging development in areas with existing infrastructure. Reduces the distance traveled by vehicles and promotes walkability and transportation efficiency. Encourages daily physical activity and improves public health.

Credit Concept
    Most people like to walk *half a mile (800 meters)* or less for regular trips such as a *daily commute*. On trips to casual destinations, they prefer to walk only five minutes or *a quarter of a mile (400 meters)* or less. Placing different kinds of destinations in close proximity achieves many documented social and environmental benefits. For instance, doubling density reduces total air pollution by 30 percent because of the shorter vehicular trips, and reduces the related *transportation's climate change effects* such as air particulate levels, and greenhouse gas emissions. In denser neighborhoods with more pedestrians and cyclists per capita, bicycle and pedestrian *injuries and deaths* tend to be fewer because of the slow vehicle speed. Because of shorter vehicular trips, the rate of car collision fatalities goes down too.

In addition, density improves *human health* in a community. In compact neighborhoods, residents bike, walk, or use transit more frequently; they are less likely to be overweight and are more physically fit. For every half-mile (800 meters) walked per day, the chance of being overweight falls around 5%. Compact development *uses existing infrastructure*, saving resources and money while more efficiently using land and preserving farmland, habitat, and open space.

This credit encourages a project location within walking distance of a variety of services ("uses") and surrounded by existing built density. The density thresholds correspond to the minimum densities needed to support fixed-rail transit (12 dwelling units or DU per acre, 30 DU per hectare) and bus transit (7 dwelling units per acre, 17.5 DU per hectare).

Two threshold types are used: 1) combining nonresidential and residential densities; and 2) separating them. Project teams have flexibility in calculating the built density in surrounding communities.

To ensure a diversity of destinations, the credit specifies uses that can and cannot count. This is to encourage occupants to combine their trips within walking distance of the project when meeting daily needs—for example, stopping at a bank on the way to the dry cleaner.

To reflect the needs of buildings devoted to housing goods (and not people), distribution centers and warehouses have different requirements for this credit. For them, proximity to transportation infrastructure is more important.

Synergies
- LTC: High-Priority Site
- LTC: Access to Quality Transit

LTC: Access to Quality Transit

Applicable to
New Construction (1-5 points)
Core and Shell (1-6 points)
Schools (1-4 points)
Retail (1-5 points)
Data Centers (1-5 points)
Warehouses and Distribution Centers (1-5 points)
Hospitality (1-5 points)
Healthcare (1-2 points)

Purpose
Reduces motor vehicle use and related greenhouse gas emissions, air pollution, and other public health and environmental harms by encouraging development in locations with multimodal transportation choices.

Credit Concept
Walkable, compact communities near transit improve the health and well-being of the community, benefit the environment, and provide alternatives to driving. Access to transit benefits people who cannot afford to own cars such as young people and the elderly.

Public transit reduces greenhouse gas emissions. Compared with low-density auto-oriented growth, developments near existing transit consume less land, reducing conversion of open spaces and farmland into built development. Investment in transit-oriented development is a proven strategy for revitalizing declining urban neighborhoods and downtowns. It brings a city about twice the economic benefit when compared with highway investment.

Transit-oriented development locations boost ridership and support transit services, and provide access to public transportation. Projects within walking distance of multiple transit routes will encourage public transportation use. Schools should be sited in a neighborhood or provide bicycle and pedestrian access to transit to reduce the number of buses required.

Synergies
- LTC: Bicycle Facilities

LTC: Bicycle Facilities

Applicable to
New Construction (1 point)
Core and Shell (1 point)
Schools (1 point)
Retail (1 point)
Data Centers (1 point)
Warehouses and Distribution Centers (1 point)
Hospitality (1 point)
Healthcare (1 point)

Purpose
Reduces vehicle use and promotes bicycling and transportation efficiency. Encourages recreational and utilitarian physical activity and improves public health.

Credit Concept
Bicycling has many global and individual benefits. About one pound (450 grams) of carbon dioxide (CO_2) emissions is avoided for every mile (1,600 meters) pedaled rather than driven. Shifting from car to bicycle use for short trips can lower risk of cardiovascular disease and extend human lives by about three to fourteen months. Bicycling as a transportation option often has popular and political support.

This credit encourages two things:
1) the provision of short- and long- term bicycle storage
2) access to a "bicycle network" (paths, trails, designated bike lanes, and slow-speed roadways)

Since regular occupants and visitors have different bicycle storage needs, long-term and short-term bicycle storage capacity is considered separately. Having to carry a bicycle into a living space discourages bicycle use. Therefore, for residential spaces, long-term storage has to be placed in an area outside individual dwelling units. Proximity to a bicycle network means building occupants can bicycle to and from the building more easily. The route destinations emphasize the role of bicycles for travel to and from work, home, and errands, as well as transit, etc.

Synergies
- LTC: Surrounding Density and Diverse Uses

LTC: Reduced Parking Footprint

Applicable to
- New Construction (1 point)
- Core and Shell (1 point)
- Schools (1 point)
- Retail (1 point)
- Data Centers (1 point)
- Warehouses and Distribution Centers (1 point)
- Hospitality (1 point)
- Healthcare (1 point)

Purpose

Minimizes land consumption, automobile dependence, rainwater runoff and other environmental harms associated with parking facilities.

Credit Concept

Inefficient parking systems generate unnecessary paved areas, create lost economic opportunities, reduce productivity, and increase carbon emissions and congestion. For example, there are approximately two to three times more parking spaces than people in the United States. Asphalt and concrete parking lots cover about 2,000 to 6,000 square miles (5,180 to 15,000 square kilometers). This is about 50% to 70% of the surface area in the average nonresidential or 35% of the surface area in the average US residential neighborhood.

Most parking spaces have impervious surfaces, and the related financial and environmental downsides to these materials. Runoff from impervious surfaces can flush contaminants into waterways and overwhelm municipal stormwater systems. Dark-colored parking lot surfaces reflect little sunlight, trap heat, contribute to the heat island effect, raise ambient air temperatures, and require more energy for cooling. Parking is expensive, and costs developers and landowners about $15,000 per space on average in the United States.

Locating projects in mixed-use, high-density areas or in places well served by transit can reduce parking demand. Other approaches include providing preferred parking for carpools and other transportation demand management strategies. To complement the use of vehicle-sharing arrangements and these alternative modes, project teams can limit vehicular parking itself by designing fewer spaces. The Transportation Planning Handbook base ratios provide a baseline to compare parking supply reductions for this credit. Projects should not provide more parking than is necessary by code. These measures ensure meaningful reductions of parking spaces.

Synergies
- LTC: Surrounding Density and Diverse Uses
- LTC: Access to Quality Transit
- LTC: Green Vehicles

LTC: Green Vehicles

Applicable to
New Construction (1 point)
Core and Shell (1 point)
Schools (1 point)
Retail (1 point)
Data Centers (1 point)
Warehouses and Distribution Centers (1 point)
Hospitality (1 point)
Healthcare (1 point)

Purpose
Promotes other alternatives to fossil fueled automobiles and reduces pollution.

Credit Concept
In 2010, about 27% of the total US greenhouse gas emissions from combustion of petroleum-based fuels came from transportation. More than 50% of those emissions came from light-duty trucks and passenger vehicles. Transportation generated 17.5% more greenhouse gas emissions in 2010 than in 1990 globally.

Conventional-fuel based vehicles create public health risks in addition to climate change effects. Idling buses generate fine particulates (FP) in their diesel exhaust, which is especially harmful for children. Diesel-powered yard tractors and idling delivery trucks emit sulfur oxides (SOx), nitrogen oxides (NOx), and particulate matter (PM).

This credit rewards regional and local infrastructure that encourages the purchase of green vehicles. Except for handicap parking spaces, preferred parking spaces are closest to a building's entrance. They can benefit building users who drive green vehicles. Third-party standards that comprehensively measure both emissions and fuel efficiency are used as criteria for green vehicles to qualify for these spaces. Projects must also provide electric charging stations or other infrastructure for alternative-fuel vehicles. Electric vehicle supply equipment can take advantage of future improvements to the utility grid, and should have effective charging speeds, which align with regional or local standards.

School projects may
1) meet emissions and green vehicle standards for their bus and non-bus vehicle fleets, or
2) address employee and visitor vehicle use, as in other rating systems.

Warehouses and distribution center projects may
1) provide electrical connections at loading dock doors, a strategy that allows truck drivers to plug into grid power to operate in-cab comfort settings and appliances rather than to idle their engines unnecessarily, and/or
2) purchase alternative-fuel yard tractors to move trailers around the facility.

Synergies
- LTC: Reduced Parking Footprint
- EAC: Demand Response
- EQP: Minimum Air Quality Performance

Chapter 7
Sustainable Sites (SS)

Overall Purpose

The Sustainable Sites (SS) category focuses on the vital relationships among ecosystems, ecosystem services, and buildings. It emphasizes integrating the site with regional and local ecosystems, restoring project site elements, and preserving the biodiversity of natural systems.

Earth's systems depend on "natural capital" such as biologically diverse coral reefs, wetlands, forests, and other ecosystems. They provide regenerative services. Per a United Nations study, about 60% of the ecosystem services that have been accessed worldwide, are currently used unsustainably or are degraded. The results are soil erosion, deforestation, disappearing rivers, drops in water table levels, and extinction of species. Sprawl and exurban development threatens natural landscapes and encroaches on the remaining farmlands, replacing and fragmenting them with hardscapes or nonnative vegetation. An area the size of Illinois, or about 34 million acres (13,759 hectares) of open space was lost to development—approximately 6,000 acres a day, or 4 acres per minute between 1982 and 2001 in the United States alone. Hardscape areas increase rainwater runoff, and overload the capacity of natural infiltration systems. Rainwater runoff carries sediment, oil, lawn fertilizers, chemicals, and other pollutants to rivers and streams; harms aquatic species and ecosystems; and contributes to **eutrophication**. For example, rainwater runoff from parking lots, roads, and other hardscapes carries about 200,000 barrels of petroleum into the Puget Sound every year per a Washington State Department of Ecology study. That is more than half of what was spilled in the 1989 Exxon Valdez accident in Alaska.

The prerequisites and credits in the SS category require project teams to complete an early site assessment, plan the locations of hardscape areas and buildings to avoid harming open space, habitat, and water bodies, and protect sensitive ecosystems. Project teams should reduce light pollution and heat island effects, minimize construction pollution, use low-impact development methods, remediate areas on the project site already in decline, and manage rainwater runoff by mimicking natural water flow patterns.

In addition to traditional approaches, the current SS category uses several new strategies, including replicating natural site hydrology (Rainwater Management credit); targeting financial support for off-site habitat protection (Site Development—Protect or Restore Habitat credit) with the help of conservation organizations; using the backlight-uplight-glare (BUG) method (Light Pollution Reduction credit); and using SR values for nonroof hardscapes (Heat Island Reduction credit) and three-year aged SRI values for roofs.

The following are some specific measures:
1) **Construction activity pollution prevention** stops soil erosion by wind or storm water runoff during construction; protects and stockpiles topsoil for reuse; avoids waterway sedimentation; and prevents airborne dust and particle matter from polluting the air.
2) **Site development that protects or restores habitat** will maximize open space and promote biodiversity via a high ratio of open space to development footprint.
3) **Storm water design with quality control and quantity control** manages storm water runoff; increases on-site infiltration; reduces impervious cover; eliminates contaminants and

storm water run-off pollution; limits disruption of natural water hydrology or natural water flows; captures and processes storm water runoff with the help of a storm water management plan and acceptable **BMPs** (best management practices).

4) **Heat island reduction for non-roof and roof** impacts the wildlife habitat, humans, and microclimate. Heat island refers to the extra thermal gradient in developed areas when compared with undeveloped areas.

5) **Light pollution reduction** lowers development impact on nocturnal environments; reduces glare and improves nighttime visibility, and reduces sky-glow and light trespass from site and building to increase night sky access.

Core Concepts
1) **Management and design of the site**
   a) Stewardship of the site
   b) Site development
   c) Light pollution reduction
   d) Pest management integration
2) **Management of storm water**
   a) Storm water quantity reduction and water quality protection
   b) Impervious surface impact

Recognition, Regulation and Incentives
Financial incentives for local, state, and federal government programs, encourage the reuse of "brownfield" sites or infill and promote water quality protection with the help of combining low impact development and smart growth.

Overall Strategies and Technologies
Note: Not all strategies and technologies have to be used simultaneously for your project.
1) Protect or restore habitat during site development.
2) Restore and/or protect open spaces. (Plan for protected areas on-site and plan for easements and protected areas off site.)
3) Manage and intercept storm water.
4) Apply cool roof technologies.
5) Reduce duration of lighting use and lighting density, and also use light fixtures that comply with dark sky requirements.

SS Prerequisite (SSP): Construction Activity Pollution Prevention

Mandatory for
- New Construction (0 points)
- Core and Shell (0 points)
- Schools (0 points)
- Retail (0 points)
- Data Centers (0 points)
- Warehouses and Distribution Centers (0 points)
- Hospitality (0 points)
- Healthcare (0 points)

Purpose
Controls airborne dust, soil erosion, and waterway sedimentation, and also reduces pollution from construction sites.

Prerequisite Concept
This prerequisite encourages project teams to reduce construction project disturbances to rainwater systems, the site itself, and neighboring properties. Local codes frequently control construction activity pollution; however, some governing agencies may not have such codes. LEED uses the US Environmental Protection Agency (EPA) **construction general permit (CGP)**, a US-based national standard, to guarantee that all projects implement **erosion and sedimentation control (ESC)** measures during construction.

Local jurisdictions typically adopt CGP and ESC measures per local weather, soils, natural waterways, and municipal rainwater systems. Projects use local codes derived from the CGP can often meet the criteria of the prerequisite.
Outside of the United States, project teams can use a local equivalent.

Synergies
- SSC: Site Development—Protect or Restore Habitat
- SSC: Rainwater Management

*SSP: Environmental Site Assessment

Mandatory for
    Schools (0 points)   Healthcare (0 points)

Purpose
    Makes sure the site is evaluated for environmental contamination, remediates the environmental contamination if the site is contaminated, and protects the health of vulnerable populations.

Prerequisite Concept
    Many sites may have been contaminated and could harm the well-being and health of future occupants. For example, groundwater or soil may have contaminants left by previous uses, or existing buildings might have hazardous materials, such as asbestos or lead. Contaminants are especially harmful to sensitive populations such as hospital patients or children. Patients have greater sensitivity to environmental stressors during treatment or illness. Children can be afflicted with cancer, acute and chronic respiratory diseases, learning disabilities, or other illnesses after exposure to toxins because they are more sensitive to toxic substances than adults. Identifying and remediating environmental contamination can help to provide a safe environment for all.

    Protect teams can use the American Society for Testing and Materials (ASTM) **environmental site assessment (ESA)** methodology for identifying and investigating a site's environmental contamination. This prerequisite encourages the use of an ESA (or local equivalent) and remediation of any confirmed site contamination to protect human health. Local assessment standards are acceptable if they are at least as stringent as ASTM Phase I and II ESA.

Synergies
- LTC: High-Priority Site

SSC: Site Assessment

Applicable to
- New Construction (1 point)
- Core and Shell (1 point)
- Schools (1 point)
- Retail (1 point)
- Data Centers (1 point)
- Warehouses and Distribution Centers (1 point)
- Hospitality (1 point)
- Healthcare (1 point)

Purpose
Evaluates site conditions to assist related design decisions and assesses sustainable options before design.

Credit Concept
A site assessment identifies **assets**, such as good solar access, healthy plant populations, and favorable climate conditions, as well as **liabilities**, such as steep slopes, pollution sources, blighted structures, unhealthy soils, and extreme climate patterns. In a word, it assesses environmental features and assists the design of a sustainable site and building.

A site assessment assists good design decisions, such as orienting buildings to take advantage of prevailing winds and solar access, placing outdoor gathering spaces near large trees or desirable water features, locating community gardens in areas with fertile soils, and optimizing the location of rainwater management features. It is part of an integrative design process, and incorporates a site's historical and ecological contexts. A good assessment performed before or during the conceptual design phase can honor a site's unique characteristics, promote occupant health, and reduce project risks and costs.

Understanding a site's climate, ambient air quality, topography, soil types, and water availability is important because these features can have significant impact on a project's ultimate performance and design.

Synergies
- IPC: Integrative Process
- LTC: Sensitive Land Protection
- LTC: Surrounding Density and Diverse Uses
- LTC: Access to Quality Transit
- LTC: Bicycle Facilities
- SSC: Site Development—Protect or Restore Habitat
- SSC: Open Space
- SSC: Rainwater Management
- SSC: Heat Island Reduction
- EAP: Minimum Energy Performance
- EAC: Optimize Energy Performance
- EAC: Renewable Energy Production
- EQC: Daylight
- EQC: Quality Views

SSC: Site Development—Protect or Restore Habitat

Applicable to
New Construction (1-2 points)
Core and Shell (1-2 points)
Schools (1-2 points)
Retail (1-2 points)
Data Centers (1-2 points)
Warehouses and Distribution Centers (1-2 points)
Hospitality (1-2 points)
Healthcare (1 point)

Purpose
Restores damaged areas to promote biodiversity and provide habitat, and protects existing natural areas.

Credit Concept
Project teams should conserve and preserve high-quality or intact native ecosystems, including their soils, hydrology, sensitive species habitat, native vegetation, and wildlife corridors. This practice helps to maintain overall ecosystem health. Project teams should situate buildings properly to protect undeveloped land.

On previously developed sites, project teams should consider restoring hydrology, soils, and native plants because they manage and filter rainwater, improve the viability of ecological communities, and fulfill other ecosystem functions. This credit encourages project teams to first consider on-site restoration.

When it is impossible to do on-site restoration, consider off-site restoration and conservation. An off-site approach can provide even greater total ecosystem benefits than on site. Providing financial support to a recognized land trust or conservation organization can be an effective tool for projects that are unable to provide contiguous, large parcels of on-site land for long-term conservation. Funds may be earmarked for activities that restore or protect habitat, such as preserving urban green space, restoring habitat that is essential to certain species, acquiring crucial land parcels, and protecting bodies of water.

Synergies
- LTC: High-Priority Site
- SSP: Construction Activity Pollution Prevention
- SSC: Site Assessment
- SSC: Open Space
- SSC: Rainwater Management
- SSC: Heat Island Reduction

SSC: Open Space

Applicable to
New Construction (1 point)
Core and Shell (1 point)
Schools (1 point)
Retail (1 point)
Data Centers (1 point)
Warehouses and Distribution Centers (1 point)
Hospitality (1 point)
Healthcare (1 point)

Purpose
Restores damaged areas to promote biodiversity and provide habitat, and protects existing natural areas.

Credit Concept
Connecting building occupants with the outdoors can improve their productivity and well-being. Open spaces also reduce heat island effect, provide increased rainwater infiltration, contribute to habitat creation and linked habitat corridors in urban areas, as well as many other many positive environmental benefits.

Features of an open space can have impact on both its environmental benefits and its use by people. For instance, paved plazas and turf fosters social interaction and encourages group activities. Vegetation in the open space contributes to the environment benefits. Open spaces could include preserved habitats with learning opportunities, vegetated roofs, community gardens, and visual interest all year long.

Synergies
- SSC: Site Development—Protect or Restore Habitat
- SSC: Rainwater Management
- SSC: Heat Island Reduction
- SSC: Site Assessment

SSC: Rainwater Management

Applicable to

New Construction (2-3 points)
Core and Shell (2-3 points)
Schools (2-3 points)
Retail (2-3 points)
Data Centers (2-3 points)
Warehouses and Distribution Centers (2-3 points)
Hospitality (2-3 points)
Healthcare (1-2 points)

Purpose

Based on undeveloped ecosystems and historical conditions in the region, replicates the natural water balance and hydrology of the site to improve water quality and reduce runoff volume.

Credit Concept

Conventional site development uses soil compaction and impervious surfaces, which often causes, loss of natural drainage patterns, loss of vegetation, and disruption of natural watersheds and hydrological systems. They disrupt the natural water flow and water balance. The rainwater management technique of a conventional site is to pipe and convey runoff into large, centralized facilities at the base of drainage areas as quickly as possible. This strategy promotes efficient drainage and prevents flooding, but increases the peak flow, temperature, volume, and duration of runoff, erodes streams, harms watersheds, and causes other ecological damage.

Low-impact development (LID) and **green infrastructure (GI)** rainwater management strategies and techniques mimic a site's natural hydrology, and improve upon the conventional approach. Project teams treat rainwater as a resource instead of a waste product. This credit uses approaches and techniques to limit the amount of impervious cover on a site, minimize disturbed areas on the project site, and then filtering, infiltrating, storing, detaining, or evaporating rainwater runoff onsite or close to its source.

Synergies

- IPC: Integrative Process
- SSC: Site Development—Protect or Restore Habitat
- SSC: Open Space
- SSC: Site Assessment
- SSC: Site Master Plan
- SSC: Heat Island Reduction
- WEP and WEC: Outdoor Water Use Reduction
- WEP and WEC: Indoor Water Use Reduction

SSC: Heat Island Reduction

Applicable to
New Construction (1-2 points)
Core and Shell (1-2 points)
Schools (1-2 points)
Retail (1-2 points)
Data Centers (1-2 points)
Warehouses and Distribution Centers (1-2 points)
Hospitality (1-2 points)
Healthcare (1 point)

Purpose
Reduces heat islands and minimizes effects on human and wildlife habitats and microclimates.

Credit Concept
Non-reflective, dark surfaces such as conventional roads, parking, walkways, roofs, and other hardscapes absorb most of the sun's warmth and radiate heat, affect microclimate, and create **heat islands**. Temperatures in urban areas are 1.8° to 5.4°F (1° to 3°C) higher than undeveloped areas and surrounding suburbs, and as much as 22°F (12°C) higher in evenings. These heat islands can be a factor in regional average warming trends. Urban heat islands contribute to 24.2% of regional warming according to a study of surface warming caused by rapid urbanization in east China. Urban heat island effects cause many negative consequences including making a habitat inhospitable for animals and plants sensitive to temperature fluctuations. Places exposed to worse ground-level pollution are often affected by heat islands too, and can harm human health. Heat islands cause larger cooling demand in the summer, meaning they require more powerful, larger air-conditioners, use more electricity, produce more greenhouse gases, increase cooling costs, and generate more pollution.

The Department of Energy's Lawrence Berkeley National Laboratory conducted a study of the metropolitan areas of Baton Rouge, Chicago, Houston, Sacramento, and Salt Lake City, and found the energy savings potential of heat island reduction measures ranged from $4 million to $15 million per year. As part of an integrated systems approach to improving building performance, measures to reduce heat islands can have a reasonable payback period. These measures include using a vegetated roof to insulate a building and extend the life of the roof or installing solar panels or shading devices.

The **solar reflectance index (SRI)** is the most effective measure of the ability of roofing materials to reject solar heat. However, in this credit, **solar reflectance (SR)** is used to measure the solar heat rejection of "non-roof" materials or components that are not considered roofing materials, such as shading devices, vegetation, and other less reflective components. SR is a more proper way to measure non-roof materials since they have more thermal mass. To measure material performance over time, this credit considers a product's three-year-aged SR values or SRI in addition to the product's initial SR value or SRI. This credit encourages project teams to adopt many strategies: incorporating high-SR or high-SRI materials such as shaded parking and vegetation; and reducing hardscape, etc. They help to alleviate heat island effects.

Synergies
- SSC: Site Development—Protect or Restore Habitat
- SSC: Open Space
- SSC: Rainwater Management
- EAP: Minimum Energy Performance
- EAC: Optimize Energy Performance

SSC: Light Pollution Reduction

Applicable to
New Construction (1 point)
Core and Shell (1 point)
Schools (1 point)
Retail (1 point)
Data Centers (1 point)
Warehouses and Distribution Centers (1 point)
Hospitality (1 point)
Healthcare (1 point)

Purpose
Reduces the consequences of development for people and wildlife, improves nighttime visibility, and increases night sky access.

Credit Concept
Artificial exterior lighting is a double-edge sword. If done correctly, exterior lighting offers comfort, extended productive hours, security, safety, aesthetics, building identification, and way finding. However, if done incorrectly it can also cause light pollution.

There are many inappropriate applications of exterior lighting, which create **light pollution**. High-angle front light causes **glare**. Artificial sky glow is caused by **uplight**. Directing light in the opposite direction of the intended area is often caused by **backlight** and can trespass onto adjacent sites.

Light pollution causes many environmental problems. When exposed to light pollution, some fauna and flora cannot adjust to seasonal variations. Wildlife species that forage or hunt at night may not be able to feed. Some migratory birds may become disoriented because they rely on stars for guidance during migration. Misuse of light can even damage human health. Overexposure to artificial lighting, especially at night, may lead to sleep disorders, impair night vision, and disrupt our melatonin production and circadian rhythms. It is a waste of both energy and money to direct light pollution into areas that do not need illumination or into the sky.

Reducing uplight, glare, and light trespass (**three forms of light pollution**) can contribute to good lighting design. This credit requires implementing an appropriate control strategy, locating lights properly on the project site, selecting the right luminaires (lamp-ballast combinations), and specifying only the necessary lighting. This credit offers two choices to give designers flexibility: a **new backlight, uplight, and glare (BUG) rating method**; and a **calculation method** (as in LEED 2009).

Responsible selection of luminaires can result in well-directed, well-shielded light and aesthetically pleasing design. Change light levels slowly to allow the human eye to adapt, and minimize direct views of lamps to avoid glare. Specify both lighting control and luminaire distribution to be efficient. Use time clocks, photocells, motion sensors, and other devices to reduce the waste of light.

Reference standard is Illuminating Engineering Society and International Dark Sky Association (IES/IDA) Model Lighting Ordinance (MLO) User Guide and IES TM-15-11, Addendum A: ies.org

Synergies
- SSC: Site Assessment
- EAP: Minimum Energy Performance
- EAC: Optimize Energy Performance

*SSC: Site Master Plan

Applicable to
Schools (1 point)

Note: *Detailed discussions have been omitted since this credit is for schools only, and is unlikely to be tested on the LEED Green Associate Exam.*

*SSC: Tenant Design and Construction Guidelines

Applicable to
Core and Shell (1 point)

Note: *Detailed discussions have been omitted since this credit is for Core and Shell only, and is unlikely to be tested on the LEED Green Associate Exam.*

*SSC: Places of Respite

Applicable to
Healthcare (1 point)

Note: *Detailed discussions have been omitted since this credit is for Healthcare only, and is unlikely to be tested on the LEED Green Associate Exam.*

*SSC: Direct Exterior Access

Applicable to
Healthcare (1 point)

Note: *Detailed discussions have been omitted since this credit is for Healthcare only, and is unlikely to be tested on the LEED Green Associate Exam.*

*SSC: Joint Use of Facilities

Applicable to
Schools (1 point)

Note: *Detailed discussions have been omitted since this credit is for Schools only, and is unlikely to be tested on the LEED Green Associate Exam.*

Chapter 8
Water Efficiency (WE)

Overall Purpose

This section looks at metering, outdoor use, indoor use, and specialized uses, and deals with water holistically. It is based on an "efficiency first" approach. Each prerequisite looks at reductions in potable water use and water efficiency alone. The WE credits recognize the use of alternative and non-potable sources of water.

Only 3% of Earth's water is fresh water, and over two-thirds of fresh water is trapped in glaciers, therefore, the creative reuse and conservation of water are important. Typically, most of water is used by the building users and then flows off-site as wastewater. In a developed country, potable water normally comes through a public water supply system from a remote site, and wastewater is piped to a remote processing plant, and then discharged into a distant water body. This pass-through system depletes freshwater aquifers and reduces streamflow in rivers, causing wells to go dry and water tables to drop. In 60% of European cities with a population over 100,000, groundwater is used too fast to be completely replenished.

Additionally, treating water for drinking and before disposal, as well as piping it to and from a building requires a significant amount of energy, which is not gauged by a building's utility meter. Water treatment and pumping consumes about 19% of all energy used in California.

Behind thermoelectric power and irrigation, buildings use 13.6% of potable water in the United States. Project teams can reuse wastewater for non-potable water needs, install water efficient fixtures, incorporate native landscapes that require no irrigation, and construct green buildings that use much less water. LEED projects saved a total of 1.2 trillion gallons (4.54 trillion liters) of water per the 2009 Green Building Market Impact Report. LEED WE credits encourage significant reduction of total water use.

Cross-Cutting Issues

The WE category includes three main components: water metering, irrigation water, and indoor water (used by appliances, fixtures, and processes, such as cooling). Several kinds of documentation are required based on the specific water-saving strategies of the project.

Site plans are used to document the locations of meters and submeters and the location and size of vegetated areas. Inside the building, floor plans show the location of indoor submeters, appliances, fixtures, and process water equipment (e.g., evaporative condensers, cooling towers). Project teams can use the same documentation for the credits in SS category.

Fixture cutsheets are manufacturers' literature, which provide specifications for each product. These must be used to document the fixtures (and appliances as applicable). The Indoor Water Use Reduction prerequisite and credit require this documentation.

Alternative water sources include rainwater harvesting, graywater reuse, municipally supplied wastewater (purple pipe water), or other reused sources. A project may earn credit in the following categories:

- WEC: Outdoor Water Use Reduction
- WEC: Indoor Water Use Reduction
- WEC: Cooling Tower Water Use
- WEC: Water Metering

However, the same water should NOT be applied to multiple credits unless it has sufficient volume to cover the demand of all the uses (e.g., toilet flushing plus irrigation demand).

Occupancy calculations are estimates based on occupant usage and are required for the Indoor Water Use Reduction prerequisite and credit. The LT and SS categories also use project occupancy calculations. You must understand how occupants are classified and counted. See WE Prerequisite Indoor Water Use Reduction for additional information.

The following are some specific measures:
1) water efficient landscaping (outdoor water reduction from a calculated midsummer baseline, that has no potable water, or irrigation use)
2) innovative wastewater technologies (indoor water that is reused when it's legal, safe, and appropriate)
3) water use reduction (indoor water)

Mnemonic
Love In Universe, or LIU (See underlined letters above.)

Core Concepts
1) Regulation of *indoor* water
   a. Save as much *indoor* potable water as possible
   b. Use water efficiently
2) *Outdoor* water
   a. Save as much *outdoor* potable water as possible
   b. Use water efficiently
3) *Process* water
   a. Reduce the need for potable water when supplying *process* water
   b. Use water efficiently

Recognition, Regulation, and Incentives
1) Recognition
   a. WaterSense is a product label sponsored by the EPA.
2) Regulation (requirements and goals)
   a. Mandatory federal water efficiency requirements per the Energy Policy Act (EPAct) of 2005 and Executive Order 13423 (2007) state that all federal facilities shall reduce water use intensity by 2% per year between 2006 and 2015 to reach a total reduction of 20%.
   b. Energy Policy Act (1992) has mandatory requirements concerning the use of water conserving plumbing fixtures in industrial, commercial, and residential buildings.
3) Financial Incentives
   a. There are local or state rebates and credits for water saving devices.

Overall Strategies and Technologies
   Note: Not *all* strategies and technologies have to be used simultaneously for your project development.
1) **Reduce *indoor* potable water demand**
   Use non-potable water, by reusing graywater, and capturing and using rainwater; reduce water use via innovative wastewater treatment; use water efficiently through waterless or high efficiency fixtures
2) **Reduce *outdoor* potable water demand**
   Use native and/or adapted and drought-tolerant plants; use non-potable water by capturing and using storm water for landscape irrigation; use water efficiently through drip-irrigation or other high efficiency technologies.
3) **Reduce potable water use for *process* water**
   Use *process* water efficiently through the use of controls and sensors; efficient management of cooling tower water; and use non-potable water.

WE Prerequisite (WEP): **Outdoor Water Use Reduction**

Mandatory for
    New Construction (0 points)
    Core and Shell (0 points)
    Schools (0 points)
    Retail (0 points)
    Data Centers (0 points)
    Warehouses and Distribution Centers (0 points)
    Hospitality (0 points)
    Healthcare (0 points)

Purpose
Reduces outdoor water use.

Prerequisite Concept
Landscape irrigation sometimes uses 30% to 70% of the water for nonagricultural use. Potable water is very valuable. For instance, in many European cities, groundwater is being used faster than it can be replenished according to the World Business Council for Sustainable Development. Use of adapted, native, and drought-tolerant plants can greatly reduce or eliminate the need for irrigation while creating good landscape design, attracting native wildlife, and integrating the building site into its surroundings. Native or adapted plants tend to require fewer chemical pesticides and fertilizers, which can be carried away in stormwater runoff and degrade water quality.

Project teams can estimate the **landscape water requirement (LWR)** by developing a water budget. This allows landscape professionals to consider the effects of irrigation system elements, planting density, plant types, and many other design variables. In landscape design, a water budget can optimize water use.

Synergies
- WEP: Building-Level Water Metering

WEP: Indoor Water Use Reduction

Mandatory for
- New Construction (0 points)
- Core and Shell (0 points)
- Schools (0 points)
- Retail (0 points)
- Data Centers (0 points)
- Warehouses and Distribution Centers (0 points)
- Hospitality (0 points)
- Healthcare (0 points)

Purpose
Reduces indoor water use.

Prerequisite Concept
A large portion of potable freshwater is used in buildings. The selection of efficient plumbing fixtures, fittings, and equipment contributes to the reduction of potable water use in buildings. Many fixtures save 20% to 50% more water than code-required levels. The US Environmental Protection Agency developed the WaterSense label to identify these efficient fixtures and ensure performance. This credit has incorporated the WaterSense label as a requirement to ensure a LEED building uses fixtures that are both high performing and water efficient.

In some buildings, water use for intensive process and appliance can be greater than the amount of water used for landscape and indoor fixtures combined. Buildings with evaporative condensers and cooling towers are good examples. Therefore, the WE category uses a separate set of requirements to address appliance and process water use.

The prerequisite addresses only the efficiency of fittings and fixtures since the WE section is designed around an "efficiency first" model; non-potable or alternative water sources that offset potable water demand are also covered in the related credit.

Synergies
- WEP: Building-Level Water Metering
- WEC: Water Metering

WEP: Building-Level Water Metering

Mandatory for
    New Construction (0 points)
    Core and Shell (0 points)
    Schools (0 points)
    Retail (0 points)
    Data Centers (0 points)
    Warehouses and Distribution Centers (0 points)
    Hospitality (0 points)
    Healthcare (0 points)

Purpose

Tracks water consumption to identify opportunities for additional water savings and supports water management.

Prerequisite Concept

Buildings may not perform as they are designed to operate. Many factors may cause discrepancies: inaccurate assumptions about occupant behavior, inadequate commissioning, flaws in energy modeling, lack of coordination in the everyday operation of the building systems, or a disconnect during the transition from construction to operations. The USGBC collects and analyzes performance data, and compares the performance of LEED buildings to identify common features among low and high performers. They share these discoveries to assist project teams in improving building performance.

There are often discrepancies between projected and actual performance of water use. The project team has to install water metering to track water use for both the project building and any related grounds. This is the first step in improving efficiency.

Synergies
- WEP: Indoor Water Use Reduction
- WEC: Water Metering
- EAP: Building-Level Energy Metering

WEC: Outdoor Water Use Reduction

Applicable to
New Construction (1-2 points)
Core and Shell (1-2 points)
Schools (1-2 points)
Retail (1-2 points)
Data Centers (1-2 points)
Warehouses and Distribution Centers (1-2 points)
Hospitality (1-2 points)
Healthcare (1 point)

Purpose
Reduces outdoor water use.

Credit Concept
See the WEP: Outdoor Water Use Reduction, Prerequisite Concept section. Using non-potable water sources and reusing potable water may further reduce potable water use.

Synergies
- SSC: Rainwater Management
- WEP: Indoor Water Use Reduction
- WEC: Cooling Tower Water Use
- WEP: Building-Level Water Metering
- WEC: Water Metering

WEC: Indoor Water Use Reduction

Applicable to
- New Construction (1-6 points)
- Core and Shell (1-6 points)
- Schools (1-7 points)
- Retail (1-7 points)
- Data Centers (1-6 points)
- Warehouses and Distribution Centers (1-6 points)
- Hospitality (1-6 points)
- Healthcare (1-7 points)

Purpose
Reduces indoor water use.

Credit Concept
See the WEP: Indoor Water Use Reduction, Prerequisite Concept section.

Synergies
- WEP: Building-Level Water Metering
- WEC: Water Metering

WEC: Cooling Tower Water Use

Applicable to
New Construction (1-2 points)
Core and Shell (1-2 points)
Schools (1-2 points)
Retail (1-2 points)
Data Centers (1-2 points)
Warehouses and Distribution Centers (1-2 points)
Hospitality (1-2 points)
Healthcare (1-2 points)

Purpose
Controls corrosion, microbes, and scale in the condenser water system, and conserves water used for the cooling tower makeup.

Credit Concept
Typically, refrigeration systems cool interior building spaces by removing heat from the air, and expelling that heat into another medium or the atmosphere. An evaporative condenser or cooling tower evaporates water to remove heat; the water changes from a liquid to a vapor when it absorbs heat. Some solids dissolve and become concentrated in the water. When the water evaporates, scale starts to deposit on evaporative condenser elements or the cooling tower. Evaporative condenser systems and cooling towers use **blowdown** to remove a portion of the water to prevent buildup of deposits. The project team then adds makeup water to replace blowdown volume and evaporative losses. Thus, a large portion of a building's total water is used in cooling towers.

Achieving target cycles of concentration is critical for reducing makeup water inputs effectively. Project teams use the number of recirculation cycles before water must be removed by blowdown to measure the water efficiency of the evaporative condenser or cooling tower.

Thousands of gallons of potable water can be saved by increasing the number of cycles during a building's peak cooling periods. Project teams can chemically analyze makeup water to find optimal cycles. In addition to input of fresh makeup water and blowdown, treating water to sequester or remove dissolved solids can also increase cycles.

Synergies
- WEP: Indoor Water Use Reduction
- WEC: Water Metering

WEC: Water Metering

Applicable to
- New Construction (1 point)
- Core and Shell (1 point)
- Schools (1 point)
- Retail (1 point)
- Data Centers (1 point)
- Warehouses and Distribution Centers (1 point)
- Hospitality (1 point)
- Healthcare (1 point)

Purpose
Tracks water use to identify opportunities for additional water savings and supports water management.

Credit Concept
Facilities managers can better measure a building's water efficiency by metering subsystem water usage. With the help of submetering the major building water systems, project teams can create independent system baselines, identify and isolate potential sources of waste, track usage against those baselines, and take corrective action. Furthermore, submetering provides the necessary data to find water savings opportunities and helps track periodic changes in water usage at a systemwide level.

WEP: Building-Level Water Metering requires a main water meter to measure the total quantity of water entering the building. This credit expands on the criteria.

Synergies
- WEP: Building-Level Water Metering
- WEP: Indoor Water Use Reduction
- WEP: Outdoor Water Use Reduction

Use Information for the WE Category
You can identify the number of occupants by **occupancy type**, i.e., two people for a one-bedroom unit, and three people for a two-bedroom unit, etc. If you do not know the occupancy (like CS projects or mixed-use projects in the early design stages where you do not know who will be the tenants), then you can use the "Default Occupancy Factors" found in the appendixes.

FTE = number of hours of occupancy/8

You can estimate the number of transients for your projects. Use a daily <u>average</u> number over a one-year period.

Use your best judgment to decide if someone should be reported as a FTE or a transient. For example, a volunteer who works at the school 4 hours each day can be considered a FTE with a value of 0.5, and an individual who attends a basketball game can be reported as a visitor.

Use a <u>1 to 1</u> male to female ratio for your projects unless they have a specific ratio. You will need to describe special ratios with a narrative.

Important Data to Memorize (The numbers in the two tables below are <u>very</u> important. Almost every LEED exam tests them.)

Fixture uses per day

Non-residential Projects					
	Water Closet	Urinal	Lavatory	Shower	Kitchen Sink
FTE (including Student FTE)					
Female	3	0	3	0.1 0 for student FTE	1
Male	1	2	3	0.1 0 for student FTE	1
Transients (Student Visitors)					
Female	0.5	0	0.5	0	0
Male	0.1	0.4	0.5	0	0
Transients (Retail Customers)					
Female	0.2	0	0.2	0	0
Male	0.1	0.1	0.2	0	0
Residential Projects					
	Water Closet	Urinal	Lavatory	Shower	Kitchen Sink
Female	5	N/A	5	1	4
Male	5	N/A	5	1	4

Note:
1) Lavatory faucets are counted as a <u>60</u> second duration per use for **residential** projects, a <u>15</u> second duration for **non-residential** projects, and a <u>12</u> second duration when equipped with autocontrol.
2) Kitchen sink faucets are counted as a <u>60</u> second duration per use for **residential** projects, and a <u>15</u> second duration for **non-residential** projects.
3) Showers are counted as a <u>480</u> second duration for **residential** projects, and a <u>300</u> second duration for **non-residential** projects.
4) This table applies to NC, CS, schools, commercial, and residential projects. I have combined several tables into one table to save you time.

Flow rates (GP**F** or GP**M** or GP**C**)

	Water Closet	Urinal	Lavatory	Shower	Kitchen Sink
Conventional (baseline)	1.6 gpf except blow-out fixture at 3.5 gpf	1.0	**Residential** 2.2gpm at 60psi **Commercial** 2.2gpm at 60 psi for private app. 0.5gpm at 60 psi for public use 0.25gpc (gallons per cycle) for metering faucets	2.5 gpm at 80psi per shower stall	2.2gpm at 60 psi
EPA Water-Sense Standard or High-efficiency	1.28 gpf	0.5 gpf	1.5 gpm for private lavatory faucets and aerators	≤2.0 gpm	1.75 gpm
HET, single-flush pressure assist	1 gpf	0.5 gpf for HEU			
HET, dual-flush (full-flush)	1.6 gpf				
HET, dual-flush (low-flush)	1.1 gpf				
HET, foam-flush	0.05 gpf				
Low-Flow	1.1 gpf	0.5 gpf	1.8 gpm	1.8 gpm	≤2.2 gpm
Ultra Low-Flow	0.8 gpm		0.5 gpm		
Non-Water Urinal or Composting Toilet		0			

Note:
1) Private app. (application) means hospital patient rooms, motel and hotel rooms, etc.
2) Commercial **pre-rinse spray valves** for food service application ≤ **1.6gpm.**
3) This table applies to **non-residential** (NC, CS, school, commercial, etc.) projects and **residential** projects. I have combined several tables into one table to save you time.
4) This table is based on the **Energy Policy Act of 1992**, **EPA WaterSense** Standard, Uniform Plumbing Code (**UPC**), and International Plumbing Code (**IPC**) standards for plumbing fixture water use. **WaterSense** is a partnership program sponsored by the **EPA**. The **EPA WaterSense** standard exceeds the IPC and UPC requirements in some cases.
5) The following fixtures shall not be included in the water use calculations for this credit, but they may be included for extra IN points for water use reduction:

- Automatic commercial ice makers, commercial dishwashers, commercial steam cookers, residential clothes washers, commercial (family-size) clothes washers, and standard or compact residential dishwashers.
- High-efficiency (HE) fixtures include single-flush gravity fed, high-efficiency toilets (HET), high-efficiency urinals (HEU), etc.

Note: HETs are available in different flush types such as single-flush gravity fed (like a conventional toilet), single-flush pressure assist, dual-flush gravity fed and dual-flush pressure assist.

Chapter 9
Energy and Atmosphere (EA)

Overall Purpose

The EA category addresses energy holistically, covering energy-efficient design strategies, energy use reduction, and renewable energy sources. Currently, a large percentage of the worldwide energy resources are from coal, oil, and natural gas. These resources are limited, nonrenewable, and emit greenhouse gases. The current energy practice involves escalating market prices due to uncertain supplies, destructive extraction processes, and national security vulnerability. This practice is not sustainable. Buildings use about 40% of the total energy, and are a major contributor to these problems

Energy efficiency design concentrates on reduction of overall energy needs through the choice of climate-appropriate building materials, glazing selection, and building orientation. Natural ventilation, passive heating and cooling, high-efficiency HVAC systems, and smart controls can reduce a building's energy use. The purchase of green power or on-site renewable energy also lowers the demand for fossil fuel energy.

An important step to ensuring high-performing buildings is the commissioning process. Get a **commissioning authority (CxA)** on board early. The CxA verifies that the design functions as intended and meets the owner's project requirements, helps to reduce wasted energy, and prevents long-term maintenance issues. The staff should understand the installed systems and their function in an operationally efficient and effective building. They should be willing to learn new methods for optimizing system performance, and must be trained so that efficient design is carried through to efficient performance.

The reduction of fossil fuel use is not limited to buildings only. The EA category recognizes this fact. Enrolling projects in a **demand response program** can help to increase electric grid efficiency.

Once enrolled in a demand response, the utility companies can decrease the building's electricity usage during peak times, which reduces the demand on the grid and the need to construct new plants. In the same way, on-site renewable energy avoids transmission losses and strain on the grid, and moves the market away from dependence on fossil fuels.

Between now and 2030, if cost-effective energy efficiency measures are employed in buildings as their lighting, cooling, heating, and other equipment is replaced, the growth in the building sector energy demand could fall from a projected 30% increase to 0% according to the American Physical Society. The EA section use credits related to designing for efficiency, reducing usage, and supplementing the energy supply with renewable energy to support the goal of reduced energy demand.

Overall, the following goals are important to the EA category:
1) Saved energy
2) Promotion of renewable energy supply
3) Fundamental commissioning for building energy systems
4) Minimum energy performance
5) Fundamental refrigerant management

6) Optimized energy performance
7) On-site renewable energy
8) Enhanced commissioning
9) Enhanced refrigerant management
10) Measurement and verification
11) Green power

Mnemonic
So, Please Forward My Files!
Oh! Oprah Enters Energy Management Group. (See underlined letters above.)

Core Concepts
1) Energy Efficiency and Demand
   - Understand energy criteria
   - Save energy
   - Measure the performance of energy
2) The Supply of Energy
   - Buy off-site renewable energy
   - Generate on-site renewable energy

Recognition, Regulation, and Incentives
1) **Recognition** through the Energy Star Program, Target Finder Rating Tool, etc.
2) **Regulation (Requirements and Goals)** See detailed discussion under each credit.
3) **Incentives**
   - Private Sector: lower risk and lower premiums for property insurance, social responsibility of corporations, and availability of money
   - Public Sector: tax rebates and credits, incentive for development, expedited plan review and permit processing, and technology-based measures.

Overall Strategies and Technologies
Note: Not **all** strategies and technologies have to be used simultaneously for your project.
1) Utilize typical energy use patterns for various building types.
2) Use statistical databases such as performance-based or prescriptive approaches.
3) Use code-based energy models.

Typical energy use pattern matrix

Building Type	Median Electrical Intensity (kWh/sf-yr)
Education	6.6
Office	11.7
Retail (except mall)	8.0
Food Sales	58.9
Food Service	28.7
Lodging	12.6

4) Reduce energy use through the use of artificial HVAC and lighting. Use energy-efficient equipment with feedbacks and controls.

5) Consider building orientation and improve building envelope performance.
6) Use Energy Star appliances and energy-efficient equipment.
7) Measure and verify building energy performance by building and operating as designed (commissioning, and continuous and retro-commissioning); monitoring performance; and improving over time (monitoring and verification).
8) Generate on-site renewable energy such as geothermal, solar PV, and wind energy.
9) Take advantage of the site's passive solar energy, natural ventilation, and passive cooling.
10) Buy off-site renewable energy.

EA Prerequisite (EAP): Fundamental Commissioning and Verification

Mandatory for

New Construction (0 points)
Core and Shell (0 points)
Schools (0 points)
Retail (0 points)
Data Centers (0 points)
Warehouses and Distribution Centers (0 points)
Hospitality (0 points)
Healthcare (0 points)

Purpose

Meets the owner's project requirements (OPR) for water, energy, durability, and indoor environmental quality; and supports the design, construction, and operation of a project to meet the OPR.

Prerequisite Concept

Commissioning (Cx) is a process to ensure that the project meets both the owner's operational needs and the design intent. An owner's objectives and goals should be the driving force for the project team. The Cx verifies that that building systems perform as intended and those objectives and goals are met.

A well-executed Cx process often leads to improved planning and coordination, fewer system deficiencies and change orders, fewer corrective actions implemented during construction, reduced energy consumption during building operation, and overall lower operating costs, as well as expressing the owner's project requirements. Better ventilation and temperature control also improves occupant health and comfort as a potential benefit of Cx. For data centers or other mission-critical facilities, Cx can reduce power or cooling system design or performance issues, the risk of information technology (IT) equipment downtime, faulty installation calibration mistakes, and discover software programming errors before the data center building is on line. Reducing the risk of information technology (IT) equipment downtime is the most significant benefit.

The qualified commissioning authority (CxA) should be brought on board early to represent the owner's needs. As a third party, the CxA can verify early on that the designs meet the OPR. During the construction phase, the CxA leads the commissioning team, and verifies that the contractors program and install the systems properly and according to the design.

Synergies

- IPC: Integrative Process
- EAC: Advanced Energy Metering
- EAC: Renewable Energy Production
- EAC: Enhanced Commissioning

EAP: Minimum Energy Performance

Mandatory for
- New Construction (0 points)
- Core and Shell (0 points)
- Schools (0 points)
- Retail (0 points)
- Data Centers (0 points)
- Warehouses and Distribution Centers (0 points)
- Hospitality (0 points)
- Healthcare (0 points)

Purpose
Achieves a minimum level of energy efficiency for the building and its systems and reduces excessive energy use and the related economic and environmental harms.

Prerequisite Concept
An integrated building design can improve thermal comfort, indoor air quality, and access to daylight as well as lower operating and maintenance costs. Project teams can use a performance or a prescriptive approach to attain such results. For a modest initial cost with a short payback period, an optimized building design can greatly reduce energy use when it includes smart operational strategies, improved mechanical system efficiency, and load reduction.

A **prescriptive approach** has a limited set of system choices with mandatory performance characteristics. It is simplified and applicable to smaller buildings, retail stores, offices, schools, and some other building types. It is ideal for projects with straightforward design, smaller budgets, and packaged equipment, and also provides energy-saving guidance for many simple buildings with standard energy systems. However, it can be inflexible because all the listed requirements must be achieved to meet the prerequisite. There are two prescriptive options available, one based on building size and the second on other factors.

A **performance approach** is more flexible, and evaluates the interactive effects of efficiency measures. It uses energy modeling to simulate the overall energy performance of a building. Project teams can then evaluate complex systems and make efficiency trade-offs among components and systems. The prescriptive options do not allow these trade-offs.

Qualified **energy modelers** can work with the design team and interpret the results of these complicated analyses to maximize the benefits. In addition to the function of a compliance check, an energy simulation can serves as a design tool if initiated early in the design process. Early energy modeling can better integrate interrelated design issues, and encourages dialogue about assumptions concerning building systems and components. This is one of the greatest benefits of early energy modeling. Information on energy use and costs can have a bigger impact on design decisions.

The USGBC has chosen **ASHRAE 90.1–2010** as the standard on which to base the requirements because it continues to push building design toward greater energy efficiency. When ASHRAE 90.1–2010 is applied instead of ASHRAE 90.1–2007, there is an average improvement of 18% across all building types according to a study by the US Department of Energy.

Synergies
- IPC: Integrative Process
- WEP: Indoor Water Use Reduction
- EAC: Optimize Energy Performance
- EAC: Demand Response
- EAC: Renewable Energy Production
- EAC: Green Power and Carbon Offsets
- EQP: Minimum Indoor Air Quality Performance

EAP: Building-Level Energy Metering

Mandatory for
    New Construction (0 points)
    Core and Shell (0 points)
    Schools (0 points)
    Retail (0 points)
    Data Centers (0 points)
    Warehouses and Distribution Centers (0 points)
    Hospitality (0 points)
    Healthcare (0 points)

Purpose

    Tracks building-level energy use, and identifies opportunities for additional energy savings and support energy management.

Prerequisite Concept

    Whole-building metering allows building operators to track energy consumption over time, analyze usage pattern variations, and develop measures for energy conservation over the life cycle of the building. Staff can then use metering to track energy savings of these conservation measures and justify additional investments with expected return periods. Metering provides building operators with detailed feedback, enabling them to operate building systems efficiently, and precisely adjust operational parameters according to the needs of changing occupancy groups.

    Buildings often perform differently from the way they are designed to operate. Even green buildings have discrepancies between projected and actual performance. Many factors can be responsible such as inaccurate assumptions regarding occupant behavior, inadequate commissioning, flaws with energy modeling, problems with the daily operation of the building systems, or the lack of coordination during the transition from construction to operations. To reduce such discrepancies, USGBC gathers and analyzes performance data, identifies common traits among high and low performers by comparing building performance across the LEED portfolio, and then shares the results to assist LEED registrants in improving their buildings' performance.

Synergies
- EAC: Advanced Energy Metering
- EAC: Enhanced Commissioning Option 1, Path 2

EAP: Fundamental Refrigerant Management

Mandatory for
- New Construction (0 points)
- Core and Shell (0 points)
- Schools (0 points)
- Retail (0 points)
- Data Centers (0 points)
- Warehouses and Distribution Centers (0 points)
- Hospitality (0 points)
- Healthcare (0 points)

Purpose
Reduces depletion of the stratospheric ozone.

Prerequisite Concept
Some refrigerants such as **chlorofluorocarbons (CFCs)** contribute to stratospheric ozone layer depletion. The thinning of this ozone layer causes many problems: damage to the marine food chain, ecological effects, such as reduced crop yields, and human health problems, such as skin cancer. The **1987 Montreal Protocol** established an international agreement to address these issues and phase out the use of CFCs and some other harmful ozone-depleting substances.

Industrialized nations that signed the Montreal Protocol before have phased out production of CFCs by December 1995, and most other countries have also achieved the goal by 2010. New CFC-based refrigeration cannot be installed for new construction projects. However, previously installed HVAC equipment may still use CFCs.

To further progress, major renovation projects must phase out CFCs from existing building equipment before building completion to meet LEED requirements. Though both CFCs and **hydrochlorofluorocarbons (HCFCs)** are factors causing ozone depletion, project teams only have to address CFCs to meet this prerequisite.

Synergies
- EAC: Enhanced Refrigerant Management
- EAC: Optimize Energy Performance

EAC: Enhanced Commissioning

Applicable to
- New Construction (2-6 points)
- Core and Shell (2-6 points)
- Schools (2-6 points)
- Retail (2-6 points)
- Data Centers (2-6 points)
- Warehouses and Distribution Centers (2-6 points)
- Hospitality (2-6 points)
- Healthcare (2-6 points)

Purpose

Expands support for the design, construction, and eventual operation of a project to meet the OPR for durability, water, energy, and indoor environmental quality.

Credit Concept

As a natural extension of the fundamental commissioning (Cx) process, **enhanced commissioning** provides owners additional verification and oversight that the building will meet their requirements and expectations during occupancy with the help of the commissioning authority (CxA). It gives the CxA the power to act as the owner's advocate and conduct in-depth reviews of the design documents, basis of design (BOD), and construction submittals. A postconstruction verification visit and training are some of the enhancements that help ongoing quality building operations and control.

The CxA can add to the Cx plan to verify that the proper tools will be provided to the building operators to manage the building's equipment effectively and efficiently. **Monitoring-based commissioning (MBCx)** supplies the operators, the CxA, and building owner with a constant flow of information to help them to identify operational issues in real time, thereby saving energy consumption, money, and time over the lifetime of the building.

Building envelope commissioning (BECx) is the other option in this prerequisite to verify and test the building's thermal envelope to achieve less energy expenditure and better building performance over its life cycle. **BECx** ensures both the passive load-defining envelope systems and the active energy-consuming systems are considered, understood, and verified. By reviewing the design documents and contractor submittals, project teams can introduce BECx early to help meet an OPR for envelope performance. These can help prevent costly problems with envelope design and construction that would be impossible or hard to deal with after construction. BECx can also improve occupant comfort through reduced solar heat gain, infiltration testing, and glare control.

Synergies
- EAP: Fundamental Commissioning and Verification
- EAC: Renewable Energy Production
- EAC: Demand Response
- EAC: Advanced Energy Metering

EAC: Optimize Energy Performance

Applicable to
   New Construction (1-18 points)
   Core and Shell (1-18 points)
   Schools (1-16 points)
   Retail (1-18 points)
   Data Centers (1-18 points)
   Warehouses and Distribution Centers (1-18 points)
   Hospitality (1-18 points)
   Healthcare (1-20 points)

Purpose
Reduces economic and environmental harms related to excessive energy use by achieving advanced levels of energy performance beyond the prerequisite standard.

Credit Concept
See the EAP: Minimum Energy Performance, Credit Concept section.

Synergies
- EAP: Minimum Energy Performance
- EAC: Renewable Energy Production

EAC: Advanced Energy Metering

Mandatory for
> New Construction (1 point)
> Core and Shell (1 point)
> Schools (1 point)
> Retail (1 point)
> Data Centers (1 point)
> Warehouses and Distribution Centers (1 point)
> Hospitality (1 point)
> Healthcare (1 point)

Purpose
> Tracks building-level and system-level energy use, identifies opportunities for additional energy savings, and supports energy management.

Credit Concept
> See the EAP: Building-Level Energy Metering, Credit Concept section.

Synergies
- EAP: Minimum Energy Performance
- EAP: Building-Level Metering
- EAC: Demand Response
- EAC: Renewable Energy Production

EAC: Demand Response

Applicable to
- New Construction (1-2 points)
- Core and Shell (1-2 points)
- Schools (1-2 points)
- Retail (1-2 points)
- Data Centers (1-2 points)
- Warehouses and Distribution Centers (1-2 points)
- Hospitality (1-2 points)
- Healthcare (1-2 points)

Purpose
Reduces greenhouse gas emissions, increases grid reliability, and increases participation in demand response programs and technologies that make distribution systems and energy generation more efficient.

Credit Concept
When there is a dramatic change in temperatures, use of heating or air-conditioning increases. In areas with clustered industrial operations or commercial buildings or urban areas, the electricity grid must respond quickly. Utilities try to keep the system at reasonable cost, operating reliably, and in balance.

Demand response (DR) strategies encourage reduction of electricity usage in peak demand hours, and help utilities optimize their supply-side delivery systems and energy generation. **Tiered demand electricity pricing** is an effective strategy. **Incentive programs** are another. They reward commercial consumers for agreeing to adjust their usage patterns when they receive an alert announcing a **DR event** (also known as a **curtailment event**) from the utility company. The alert is typically sent to the building automation system or the building's operator. DR programs specify the time frames for possible DR occurrence and set a maximum number of potential events.

DR helps utility companies avoid building additional distribution stations, transmission lines, and power generation facilities by reducing total demand for electricity. As a result, some of the environmental effects of energy consumption and infrastructure are avoided. DR also contributes to balance the supply of renewable energy sources. For instance, at night or on calm days, when renewable sources such as solar and wind are less available, grid operators have to convince energy users to lower demand or find additional generation sources. DR achieves the former by reducing the need for nonrenewable backup generation and balancing system-wide usage.

Synergies
- EAP: Fundamental Commissioning and Verification

EAC: **Renewable Energy Production**

Applicable to
- New Construction (1-3 points)
- Core and Shell (1-3 points)
- Schools (1-3 points)
- Retail (1-3 points)
- Data Centers (1-3 points)
- Warehouses and Distribution Centers (1-3 points)
- Hospitality (1-3 points)
- Healthcare (1-3 points)

Purpose
Increases self-supply of renewable energy, and reduces the economic environmental harms related to fossil fuel energy.

Credit Concept
Renewable energy can offer local environmental benefits, and reduce carbon emissions and air pollution. Some renewable energy systems use waste; others capture sunlight or wind. On-site renewable energy production protects projects from reliance on the grid and energy price volatility while reducing energy transmission loss. It also reduces a country's reliance on and demand for imported energy.

Synergies
- EAC: Advanced Energy Metering
- EAC: Green Power and Carbon Offsets

EAC: Enhanced Refrigerant Management

Applicable to
- New Construction (1 point)
- Core and Shell (1 point)
- Schools (1 point)
- Retail (1 point)
- Data Centers (1 point)
- Warehouses and Distribution Centers (1 point)
- Hospitality (1 point)
- Healthcare (1 point)

Purpose
Minimizes direct contributions to climate change, reduces depletion of ozone layer, and supports early compliance with the Montreal Protocol.

Credit Concept
This credit deals with global warming potential (GWP) and ozone depletion potential (ODP), the two main environmental threats caused by refrigerants. Hydrochlorofluorocarbons (HCFCs), chlorofluorocarbons (CFCs), and some other substances used in refrigerants are factors that cause the depletion of the stratospheric ozone layer. Refrigerants also have a large impact on global climate change compared with other greenhouse gases. For instance, the impact of HCFC-22 on warming is 1,780 times the potency of an equal amount of carbon dioxide.

Nevertheless, project teams also need to make trade-offs between energy use and the above concerns. HFC-410A and some other alternatives to HCFC and CFC refrigerants, have a lower GWP but may require more energy, which affects climate. Variable refrigerant flow (VRF) systems may have a higher refrigerant charge but improve energy efficiency.

Cautious consideration of the refrigerant criteria of appliances and energy systems can reduce operating cost and improve performance. Refrigerants vary in material compatibility, operating pressure, toxicity, and flammability. Material compatibility and operating pressure are especially important factors to consider when replacing existing equipment's refrigerants.

The calculation of refrigerant impact deals with the total effect of each refrigerant's GWP and ODP as well as these interrelated factors.

Synergies
- EAC: Optimize Energy Performance

EAC: Green Power and Carbon Offsets

Applicable to
    New Construction (1-2 points)
    Core and Shell (1-2 points)
    Schools (1-2 points)
    Retail (1-2 points)
    Data Centers (1-2 points)
    Warehouses and Distribution Centers (1-2 points)
    Hospitality (1-2 points)
    Healthcare (1-2 points)

Purpose
Promotes the use of carbon mitigation projects, renewable energy grid-source technologies, and the reduction of greenhouse gas emissions.

Credit Concept
The market can encourage utility companies and energy generators to help address climate change by developing clean energy sources. Buildings using nonrenewable power can purchase **renewable energy certificates (RECs)** to create renewable energy market demand. **Carbon offsets** let companies or buildings fund energy efficiency projects, methane abatement, reforestation or land-use changes, and other activities to remove carbon from the atmosphere or decrease carbon emissions.

Synergies
- EAP: Minimum Energy Performance
- EAC: Optimize Energy Performance
- EAC: Renewable Energy Production

Some Important Concepts

Process energy includes energy required for refrigeration and kitchen cooking, laundry washing and drying, elevators and escalators, computers, office and general miscellaneous equipment, lighting not included in the lighting power allowance (such as lighting that is part of the medical equipment), other uses like water pumps, etc.

Regulated (non-process) energy includes energy required for HVAC; exhaust fans and hoods; lighting for interiors, surface parking, garage parking, building façade, and grounds; space heating; service water heating; etc.

To achieve EA Credit Optimize Energy Performance, you should use the same process loads for your proposed building performance rating and the baseline performance building rating. You can use the Exceptional Calculation Method (ANSI/ASHRAE/IESNA 90.1-2007 G2.5) to record measures of reducing process loads. Include the assumptions made for both your proposed design and base building, as well as related supporting empirical and theoretical information in your documentation of process load energy savings.

Combined heat and power systems (**CHP**) capture the heat that would have been wasted in the process of generating electricity via fossil fuel. They are much more efficient than separate thermal systems and central power plants. They reduce peak demand, generate fewer emissions, reduce loss in electricity transmission and distribution, and release electrical grid capacity for other uses.

Chapter 10
Materials and Resources (MR)

Overall Purpose

The MR credit category emphasizes minimizing the impacts such as the embodied energy that is associated with the maintenance, processing, extraction, transport, and disposal of building materials. The criteria are set up to encourage resource efficiency and support a **life-cycle approach,** which in turn improves performance. Each criterion recognizes a particular action that is part of a life-cycle approach to embodied impact reduction.

The Hierarchy of Waste Management

About 25% of the total waste stream in the European Union and about 40% of the total solid waste stream in the United States is construction and demolition waste. **Source reduction, reuse, recycling, and waste to energy** are the four preferred strategies for reducing waste in the **solid waste management hierarchy** for the US Environmental Protection Agency (EPA) ranks. The MR section directly covers each of these recommended strategies.

Source reduction prevents environmental damages in a material's entire life cycle, from the supply chain, to actual use, and finally to recycling and waste disposal. It is at the top of the hierarchy in regards to efficiency. Source reduction promotes designing and prefabrication of dimensional construction materials and other innovative construction strategies to minimize inefficiencies and material cutoffs.

Building and material **reuse** prevents environmental harms caused by the manufacturing process, and is the second most effective strategy. Production and transportation of new materials are required to replace existing materials with new ones. The process generates greenhouse gases and takes many years of increased building efficiency to be offset. LEED v4 is more flexibile and rewards all material reuse in a project, including both in situ and from off site, as well as building reuse and salvaging strategies. Rewarding material reuse is a consistent LEED strategy.

Recycling is the most common practice to divert waste from landfills. Traditionally, most waste is sent to a landfill and with time is becoming more and more of an unsustainable solution. Landfill space is being used up in urban areas, requires more land elsewhere, and raises the waste transportation costs. Recycling technology innovations improve processing and sorting, keep the materials in the production stream for longer, and supply raw material to secondary markets.

Secondary markets only exist for some materials. The conversion of energy is its next most beneficial use. Many countries use a waste-to-energy solution to alleviate the burden on landfills. In Saudi Arabia, Sweden, and some other countries, there are far more waste-to-energy facilities than landfills. Waste-to-energy can be a feasible alternative to energy production by extracting fossil fuels when air quality control measures are strict.

Overall, LEED projects have diverted more than 80 million tons (72.6 million tonnes) of waste from landfills, and are expected to divert 540 million tons (489.9 million tonnes) by 2030. LEED projects in Seattle diverted 175,000 tons (158,757.3 tonnes), which is an average of 90% of their construction waste from the landfill from 2000 to 2011. The result would be astounding if all newly constructed buildings can achieve the 90% diversion rate. Construction debris is a resource instead of waste.

Life-Cycle Assessment in LEED

Using MR category credits, LEED has created a cycle of end user demand and industry distribution of environmentally preferable products, and started to transform the building products market. LEED has generated demand for increasingly sustainable products, and designers, manufacturers, and suppliers are responding. LEED has measurably increased the supply of sustainable materials, such as bio-based materials, increased recycled content, and harvested wood. Some MR credits promote the use of products that meet specific criteria. However, some products that have different sustainable attributes are hard to compare. For instance, solid wood cabinets made from local timber versus cabinets bound together in resin and made of wheat husks sourced from no local areas. **Life-cycle assessment (LCA)** offers a more complete picture of products and materials, allows project teams to make more informed choices that will have greater overall benefit for the human health, environmental, and communities, while promoting innovation and encouraging manufacturers to improve their products.

According to ISO 14040 International Standard, Environmental management, Life cycle assessment, principles and framework (Geneva, Switzerland: International Organization for Standardization, 2006), **LCA** is a "compilation and evaluation of the inputs and outputs and the potential environmental impacts of a product system throughout its life cycle." Project teams examine the entire life cycle of a building or product, identify the constituents and processes, and assess their environmental effects both upstream and downstream, from the point of raw materials extraction or manufacture, to transportation, use, maintenance, and end of life. LCA is also called **"cradle to grave."** Going one step further, **"cradle to cradle"** emphasizes reuse and recycling instead of disposal.

LCA started with carbon accounting models in the 1960s. LCA practices and standards have been refined and developed since then. Regulators, specifiers, manufacturers, and consumers in many fields have been using life-cycle information to improve their product environmental profiles and product selections in Europe and some other places. However, the United States lacked the tools and data that support LCA until recently. Now more and more manufacturers are ready to detail and publicly release the environmental profiles of their products, and programs are available to help users understand the results and assist this effort.

LEED strives to speed up the use of LCA-based decision-making and LCA tools, so as to improve the quality of databases and stimulate market transformation. Because of the limitations of LCA for dealing with the ecosystem consequences and human health of raw material extraction, LEED uses different, complementary approaches to LCA in credits covering these topics.

Cross-Cutting Issues
Required Products and Materials
The MR credit category includes the portions of the building or the whole building that are being renovated or constructed. Unless otherwise noted, project teams typically exclude portions of an existing building that are not in the renovation scope from MR documentation. See the minimum program requirements (MPR) for information on additions.

Qualifying Products and Exclusions
The MR section covers "permanently installed building products," which means materials and products that are attached to a building or that create the building. These include enclosure and structural elements, framing, interior walls, installed finishes, cabinets and casework, doors, and

roofs, etc. Most of these materials belong to Construction Specifications Institute (CSI) 2012 MasterFormat Divisions 3-10, 31, and 32. Some products covered by MR credits may be outside these divisions.

Project teams do not have to include furniture in credit calculations. Nevertheless, if project teams include furniture in MR credit calculations, they must include all furniture consistently in all cost-based credits.

In previous versions of LEED, the USGBC excluded all mechanical, plumbing, and electrical equipment (MEP), part of CSI MasterFormat divisions 11, 21-28, and other specialty divisions from MR credits. In LEED v4, some specific products that are "passive" parts of the system (not part of the active portions of the system) may be included. This allows optional assessment of ducts, duct insulation, conduit, lamp housings, piping, pipe insulation, plumbing fixtures, showerheads, and faucets. If project teams include them in credit calculations, they must include them consistently in relevant MR credits. Nevertheless, unlike furniture, if project teams include some of these products in credit calculations, they do NOT have to include all products of that type. For instance, if project teams include the cost of ducts in the MR calculations for recycled content, they do NOT have to include the cost of ducts that do not meet the credit requirement in the denominator or numerator of the credit calculation. Nevertheless, cost-based credits (all Building Product Disclosure and Optimization credits) calculations must have the same denominator.

Process equipment, fire suppression systems, elevators, escalators, and other special equipment, is not included in the credit calculations. Concrete formwork and other products purchased for temporary use on the project are also excluded.

For healthcare projects, the MR Credit Medical Furniture and Furnishings includes all freestanding medical furnishings and furniture. To avoid double-counting, freestanding furniture items counted in this credit cannot be included in any Building Product Disclosure and Optimization credit. Built-in millwork, casework and other permanently installed items must be counted in the Building Product Disclosure and Optimization credits, not MR Credit Medical Furniture and Furnishings.

Defining a Product
Some credit calculations in this category are based on the number of products instead of product cost. In these cases, a "permanently installed building product" or a "product" is defined by its function. A product includes the necessary services and physical components to serve its function. For similar products within a specification, each is counted as a separate product. Here are a few samples.

Products shipped to the project site ready for installation-
- Concrete masonry units, wallboard, and metal studs are all separate products.
- For wallboard, the binder, the gypsum, and backing are all required for its function, so each ingredient is not a separate product.

Products shipped to the project site as a component or ingredient used in a site-assembled product-
- Since each component in concrete (aggregate, admixture, and cement) serves a different function, each component is considered a separate product.

Similar products from the same manufacturer with reconfigurations or aesthetic variations versus similar products from the same manufacturer with different formulations-

- Since paint types of distinct gloss levels, such as gloss, semi-gloss, and flat paint, are specified to serve a specific function, such as water resistance, they are separate products. Since different colors of the same paint serve the same function, they are not separate products.
- Since carpets of different pile heights are used for different kinds of foot traffic, they are separate products. Since the carpets in the same product line but in a different color serve the same function, they are not separate products.
- Since side chairs and desk chairs in the same product line serve different functions, they are different products. Since two side chairs with only different aesthetic aspects, such as the presence of arms, serve the same function, they are not different products.

Determining Product Cost
Product and material cost includes all contractor expenses and taxes to ship the material to the project site but does *not* include any cost for equipment and labor necessary for installation after the material is shipped to the site.

Use either the default materials cost or the actual materials cost to calculate the total materials cost of a project.

Default materials cost calculates 45% of the total construction costs as an option to determine the total materials cost. It can replace the actual cost for most products and materials, as listed above. If optional products and materials, such as MEP items and furniture are included, project teams should add the actual value of those items to the default value for all other materials and products.

Actual materials cost is the cost of all the materials used on the job site, including delivery and taxes, but excluding labor.

Location Valuation Factor
Some MR category credits include a location valuation factor which adds value to locally produced materials and products. The purpose is to encourage the purchase of products and support the local economy. Materials and products manufactured, extracted, and purchased within 100 miles (160 kilometers) of the project are valued at two times their cost.

To qualify for the location valuation factor, a product must meet two conditions:
1) All manufacture, extraction, and purchase (including distribution) of the product and its materials must occur within 100 miles (160 kilometers) of the project.
2) The product (or portion of an assembled product) have to meet a minimum of one of the sustainable criteria (e.g., recycled content, FSC certification) listed in the credit.

Materials and products that do not meet the location criteria but do meet a minimum of one of the sustainability criteria are valued at one times their cost.

The distance should be measured with a straight line instead of actual travel distance. The location of the purchase transaction is considered the point of purchase. The location of product distribution is considered the point of purchase for transactions that do not occur in person such as online purchase.

See MRC: Building Product Disclosure and Optimization, Sourcing of Raw Materials, for the location valuation factor of reused and salvaged materials.

Determining Material Contributions of an Assembly

Many MR category sustainability criteria are applicable for the entire product, as is the case for programs and product certifications. Nevertheless, some criteria are only applicable for a portion of the product. The portion of the product that helps to earn points for the credit could be either the percentage of qualifying components permanently or mechanically attached together or a percentage of a homogeneous material. In either situation, the project teams should use *weight* to calculate the contributing value. Assemblies (parts permanently or mechanically attached together) include demountable partition walls, premade window assemblies, office chairs, doors, etc. Homogeneous materials include ceiling tiles, composite flooring, and rubber wall base, etc.

Based on weight, project teams can calculate the value that contributes to the credit as the percentage of the component or material meeting the criteria, multiplied by the total product cost.

Product value ($) = Total product cost ($) x (%) product component by weight x (%) meeting sustainable criteria

Below is some useful information.
The MR category has many purposes including the following:
1) <u>m</u>inimize material use (**reduce**)
2) <u>e</u>nvironmentally friendly materials
3) <u>w</u>aste management and reduction (**recycle**)
4) <u>s</u>torage and collection of recyclables
5) <u>b</u>uilding **reuse** (maintain existing walls, floors and roof)
6) <u>b</u>uilding reuse (maintain interior nonstructural elements)
7) <u>c</u>onstruction waste management
8) <u>m</u>aterials reuse
9) <u>r</u>ecycled content
10) <u>r</u>egional materials
11) <u>r</u>apidly renewable materials
12) <u>c</u>ertified wood

Mnemonic
<u>M</u>y <u>e</u>ffortless <u>w</u>ork at <u>SBBC</u>
<u>MR</u> <u>R</u>egan <u>R</u>ay <u>C</u>arter (See underlined letters above.)

Core Concepts
1) Manage waste
   - Reduce waste
   - Reuse and divert waste
2) Reduce and reuse materials
   - Reduce materials used
   - Reuse of building and materials
   - Choose rapidly renewable materials

Recognition, Regulation, and Incentives
1) Recognition
   - Cradle to Cradle, Green Seal and other product certifications
2) Regulation (requirements and goals)
   - Rare
   - Internal policy for supply chain and materials management in some organizations
3) Financial Incentives
   - Recycling incentive

Overall Strategies and Technologies
   Note: Not **all** strategies and technologies have to be used simultaneously for your project.
1) **Reduce** waste.
2) Purchase sustainable materials.
3) **Reuse** and divert waste. **Recycle** solid waste and demolition waste.
4) Reduce life cycle impact.
5) Reduce demand for materials. (Implement in design and construction and use new technologies.)
6) Reuse all or a portion of the existing building.
7) Reuse materials. (Use refurbished, salvaged, and reclaimed materials; purchase refurbished and reclaimed materials.)
8) Use **rapidly renewable materials** such as wool carpeting, cork flooring, sunflower seed board panels, linoleum flooring, bamboo flooring, and cotton batt insulation.
   Mnemonic: WC on SLAB (See underlined letters above.)
9) Choose materials with a reduced life cycle impact such as regional materials, certified wood, and materials containing pre- and post-consumer recycled content.

MR Prerequisite (MRP): Storage and Collection of Recyclables

Mandatory for
New Construction (0 points)
Core and Shell (0 points)
Schools (0 points)
Retail (0 points)
Data Centers (0 points)
Warehouses and Distribution Centers (0 points)
Hospitality (0 points)
Healthcare (0 points)

Purpose
Reduces the waste generated by building users, and diverts solid waste from landfills.

Prerequisite Concept
Waste disposal remains a major environmental burden on ecosystems and communities. In the United States, approximately 69% of total municipal solid waste is from paper, food, metal, glass, and plastic, all of which are recyclable. Building owners can divert recyclables from landfills, help convert them into new products, and reduce hauling costs and demand for virgin materials.

A lack of convenient, physical spaces for recycling is a common problem. Adding recycling infrastructure in the design stage promotes successful recycling in the operations stage. Accessible and well-designed waste management infrastructure at proper locations helps building occupants form habits of recycling.

More and more old computers, keyboards, cameras, and printers are becoming electronic waste (e-waste). This is a growing environmental concern. It is important to identify recycling facilities, storage areas, and haulers that are able to process e-waste. Fluorescent lamps, batteries, and other e-waste materials are hazardous and require a separate procedure when compared with regular recycled materials such as cardboard, glass, plastic, metals, and paper. This prerequisite requires project teams to develop waste management infrastructure for a minimum of two hazardous waste streams.

Synergies
- MRP: PBT Source Reduction—Mercury (Healthcare)

MRP: Construction and Demolition Waste Management Planning

Mandatory for
    New Construction (0 points)
    Core and Shell (0 points)
    Schools (0 points)
    Retail (0 points)
    Data Centers (0 points)
    Warehouses and Distribution Centers (0 points)
    Hospitality (0 points)
    Healthcare (0 points)

Purpose
    Recovers, reuses, and recycles materials, and also diverts demolition and construction waste from landfills and incineration facilities.

Prerequisite Concept
    A large portion of the worldwide waste is construction waste. Approximately 170 million tons of demolition and construction waste was produced, including 61% from nonresidential construction projects, in the United States in 2003 according to the US Environmental Protection Agency (EPA). About 510 million metric tonnes of construction waste is produced each year by nations in the European Union per the European Commission. Diverting this waste from landfills promotes recycling, prevents water and ground pollution, and allows for active use of the material longer.

    Different places may have completely different waste management services. Project teams should first identify local haulers, technologies, and facilities. Arrange for **construction waste management (CWM)** early on, ideally before construction to allow time to find the most effective waste diversion strategies available. **Reuse, recycle, donate, and salvage** are some typical strategies; source separation and source reduction are also effective and feasible. **Source separation** separates waste to ensure delivery to the correct facility. **Source reduction** reduces project waste through modular construction, prefabrication, or designing with standard material sizes or lengths.

    Project teams will have more time for identifying appropriate strategies, planning and coordination, and developing contractual agreements if they develop a CWM plan early in the design process. Educating site workers, project team members, and waste haulers contributes to success by ensuring everyone follows the plan and actually diverts the proper materials from landfills and incinerators. A well-devised CWM plan can contribute to selling high valued scrap materials, decreasing tipping fees, or identifying materials for reuse, minimizing cost, and maximizing return.

Synergies
- MRC: Construction and Demolition Waste Management

*MRP: PBT Source Reduction—Mercury

Applicable to
    Healthcare (0 points)

Note: *Detailed discussions have been omitted since this prerequisite is for Healthcare only, and is unlikely to be tested on the LEED Green Associate Exam.*

MRC: Building Life-Cycle Impact Reduction

Applicable to
    New Construction (2-5 points)
    Core and Shell (2-6 points)
    Schools (2-5 points)
    Retail (2-5 points)
    Data Centers (2-5 points)
    Warehouses and Distribution Centers (2-5 points)
    Hospitality (2-5 points)
    Healthcare (2-5 points)

Purpose
    Takes full advantage of the environmental performance of materials and products, and promotes adaptive reuse.

Credit Concept
    Buildings have global, regional, and local environmental impacts over their life cycles. Some involve the extraction, harvest, manufacture, and transportation of materials; others occur during operations and construction; still others happen at disposal and demolition. **A life-cycle assessment (LCA)** scrutinizes as many of these environmental impacts as possible. This credit recognizes several strategies for reducing these environmental harms such as reducing a building's environmental footprint through LCA, reusing building components, or restoring existing buildings.

    Rehabilitating blighted buildings, preserving historic buildings, and restoring existing structures decreases waste and the energy use related to construction demolition. Building reuse offers environmental savings over demolition and new construction at all times according to a report by the National Trust for Historic Preservation, titled *The Greenest Building: Quantifying the Environmental Value of Building Reuse*. New energy efficient buildings will not compensate for the climate change impacts of their construction for at least 10 years and maybe 80 years. Restoring existing buildings preserves a site's aesthetic, historical, and cultural values, and repurposing or reusing stone, wood, steel, brick, or other materials from off site can be a sustainable and cost-effective strategy.

    A cradle-to-grave LCA allows project teams to understand the environmental consequences and cumulative energy used through all phases of a newly constructed building's life. A quantitative, comprehensive analysis helps the project team to select materials best fit the project's needs in the building's life cycle. LCA is a cost-effective design tool, which can reduce the amount of materials used (**"dematerialization"**) and the related environmental impacts. An LCA also enables the project team to recognize the trade-offs of energy performance and material selection and attain their balance. For instance, high thermal mass costs money but can reduce a building's peak energy demands; an LCA can quantify the environmental impacts related to the additional materials used and allows the project team to compare the impacts with the energy performance benefits and make informed design decisions. By reviewing how materials interact within the whole enclosure and structure instead of just reviewing them as individual items, the project team can reduce overall environmental impacts and gain a larger perspective in the long term.

    The whole-building LCA approach considers many of such effects. These effects include eutrophication, acidification of water sources and land, depletion of nonrenewable energy sources, stratospheric ozone depletion, global warming potential, and formation of tropospheric ozone. Those are just some of the most well-understood, measurable, and common environmental effects that LCA tools assess. Existing LCA tools cannot accurately measure land-use issues, ecological, and human health; nevertheless, those effects are addressed under other MR credits since they are also important to a lifecycle approach to materials.

Synergies
- LTC: High-Priority Site
- MRP: Construction and Demolition Waste Management Planning
- MRC: Construction and Demolition Waste Management
- MRC: Building Product Disclosure and Optimization—Sourcing of Raw Materials
- MRC: Building Product Disclosure and Optimization—Environmental Product Declarations
- MRC: Building Product Disclosure and Optimization—Material Ingredients

MRC: Building Product Disclosure and Optimization— Environmental Product Declarations

Applicable to
- New Construction (1-2 points)
- Core and Shell (1-2 points)
- Schools (1-2 points)
- Retail (1-2 points)
- Data Centers (1-2 points)
- Warehouses and Distribution Centers (1-2 points)
- Hospitality (1-2 points)
- Healthcare (1-2 points)

Purpose

Encourages selection of manufacturers and products with verified improved environmental life-cycle impacts. Rewards project teams for using materials and products with life-cycle information that have social, economical, and environmental preferred life-cycle impacts.

Credit Concept

This credit acknowledges the selection of products with well-known environmental impacts because of industry standard reporting protocols and life-cycle information. **Environmental product declarations (EPDs)** are a standardized method of conveying the environmental effects related to a system or product's energy use, waste generation, chemical makeup, raw material extraction, and emissions to air, soil, and water. Even though many EPD programs are available, the credit requires that EPDs originate from program operators who abide by the **International Organization for Standardization (ISO)** standards, the internationally recognized norm for EPDs. With EPDs, project teams can evaluate and compare similar products more accurately, and make better decisions on material selection.

This credit will acknowledge the most advanced disclosures available as EPDs to become more and more common. Project teams should give preference to products with EPDs. This rewards the transition from a "single-attribute" approach to more comprehensive reporting, and encourages manufacturers with products that are less harmful to the environment.

To encourage both leadership and initial first steps in life-cycle information disclosure, this credit has different compliance paths and options. The purpose of this credit is to help transform the market for building materials and products with life-cycle information, and reward the manufacturers with verified environmental performance.

Synergies
- MRC: Building Product Disclosure and Optimization—Sourcing of Raw Materials
- MRC: Building Product Disclosure and Optimization—Material Ingredients
- MRC: Building Life-Cycle Impact Reduction

MRC: **Building Product Disclosure and Optimization—Sourcing of Raw Materials**

Applicable to
New Construction (1-2 points)
Core and Shell (1-2 points)
Schools (1-2 points)
Retail (1-2 points)
Data Centers (1-2 points)
Warehouses and Distribution Centers (1-2 points)
Hospitality (1-2 points)
Healthcare (1-2 points)

Purpose
Encourages project teams to select products confirmed to have been sourced or extracted in a responsible manner. To reward the use of materials and products with life cycle information which have social, economical, and environmental preferable life cycle effects.

Credit Concept
Extracting raw material has a direct environmental effect on ecosystems. For instance, conventional logging is the major cause for deforestation in subtropical Asia and Latin America, depleting over 70% of the resources; mining operations account for another 18% of the world's deforestation. In addition to deforestation, unmanaged extraction practices can also cause releases of toxic chemical, habitat loss, threats to rare and endangered species, degradation of water sources, and the infringement of indigenous peoples' rights.

This credit promotes the use of responsibly extracted and sourced materials through demonstration and reporting of responsible extraction practices. **Corporate sustainability reports (CSRs)**, based on generally accepted standards and frameworks, can identify sources of raw material extraction and provide information on product supply chains. CSRs have gained popularity among many businesses, from product manufacturers to retail organizations. CSRs provide frameworks that enable transparency and environmental effects to be improved, evaluated, and compared with other companies as sustainability goals become more and more important.

This credit also encourages project teams to select recycled and reused materials, and reduce raw material usage besides seeking the responsible sourcing of virgin materials. Teams can also abide by leadership performance standards and certifications that promote local sourcing. To acknowledge the rapidly transforming marketplace conditions for material and product reporting, this credit uses an additional "USGBC-approved program" criterion to recognize any upcoming leadership certification programs.

This credit promotes environmental impact reductions that have positive effects on the sources of project materials and go beyond the individual project by increasing the demand for transparency in quarrying, mining, forestry, agriculture, and other industries.

Synergies
- MRC: Building Life Cycle Impact Reduction
- MRC: Building Product Disclosure and Optimization—Environmental Product Declarations
- MRC: Building Product Disclosure and Optimization—Material Ingredients

MRC: Building Product Disclosure and Optimization—Material Ingredients

Applicable to

New Construction (1-2 points)
Core and Shell (1-2 points)
Schools (1-2 points)
Retail (1-2 points)
Data Centers (1-2 points)
Warehouses and Distribution Centers (1-2 points)
Hospitality (1-2 points)
Healthcare (1-2 points)

Purpose

Encourages raw material manufacturers to produce products confirmed of having improved life-cycle effects. Rewards the use of materials and products with life cycle information and that have socially, economically, and environmentally preferable life cycle effects. To encourage project teams to select products confirmed to minimize the generation and use of harmful substances, and products whose chemical ingredients are inventoried via an established methodology.

Credit Concept

An average building user has little information about the components of the building that he uses every day. Since disclosure data is hard to obtain, even project planners do not have adequate knowledge about construction materials in order to make the selection. Even though the regulations exist for some toxic chemicals, 96% of the approximately 85,000 chemicals used in the United States have not been screened for potential health impacts.

Many building materials and products contain **persistent bioaccumulative and toxic chemicals (PBTs)** and **persistent organic pollutants (POPs)**. PBTs can cause damage even in very small amount, build up in humans and other life forms high on the food chain, and remain in the environment. PBTs discharged during the use, manufacture, or disposal of a product can jeopardize the health of animals and plants far away. Even less is understood about which chemicals are potential neurotoxicants, mutagens, carcinogens, or developmental toxicants.

By encouraging green chemistry and complying with the preventive principle, this credit rewards project teams for avoiding products with possible detrimental chemicals, and stimulates innovation in materials from manufacturers. According to *Report of the United Nations Conference on Environment and Development,* "Where there are threats of serious or irreversible damage, lack of full scientific certainty shall not be used as a reason for postponing cost-effective measures to prevent environmental degradation."

See the following link:
http://www.un.org/documents/ga/conf151/aconf15126-1annex1.htm

Green Chemistry is the design of chemical processes and products to eliminate or reduce generation and use of harmful substances. Green chemistry research is attempting to develop safer substitutes, prioritize chemicals, create "green lists", and replace "red lists." Green firms are also establishing credible assurance and monitoring programs, developing corporate policies including the preventive principle, and improving relationships with suppliers.

This credit is to encourage manufacturers to disclose information about their product ingredients, and enable project teams to make better-informed choices. The programs in this credit use hazard assessment approaches to assess multiple environmental and human health endpoints in a way that is more detailed than most life-cycle assessments. Project teams may provide manufacturers' reports to prove responsible product selection, or use specified programs to ensure the absence of materials that are of concern.

Synergies
- MRC: Building Life-Cycle Impact Reduction
- MRC: Building Product Disclosure and Optimization—Environmental Product Declarations
- MRC: Building Product Disclosure and Optimization—Sourcing of Raw Materials

MRC: **PBT Source Reduction—Mercury**

Applicable to
  Healthcare (1 point)

Note: *Detailed discussions have been omitted since this credit is for Healthcare only, and is unlikely to be tested on the LEED Green Associate Exam.*

MRC: **PBT Source Reduction—Lead, Cadmium, and Copper**

Applicable to
  Healthcare (2 points)

Note: *Detailed discussions have been omitted since this credit is for Healthcare only, and is unlikely to be tested on the LEED Green Associate Exam.*

MRC: **Furniture and Medical Furnishings**

Applicable to
  Healthcare (1-2 points)

Note: *Detailed discussions have been omitted since this credit is for Healthcare only, and is unlikely to be tested on the LEED Green Associate Exam.*

MRC: **Design for Flexibility**

Applicable to
  Healthcare (1 point)

Note: *Detailed discussions have been omitted since this credit is for Healthcare only, and is unlikely to be tested on the LEED Green Associate Exam.*

MRC: Demolition Waste Management

Applicable to
New Construction (1-2 points)
Core and Shell (1-2 points)
Schools (1-2 points)
Retail (1-2 points)
Data Centers (1-2 points)
Warehouses and Distribution Centers (1-2 points)
Hospitality (1-2 points)
Healthcare (1-2 points)

Purpose
Utilizes material reuse, recover, and recycle. Diverts demolition and construction waste away from landfills and incineration facilities.

Credit Concept
With better reuse and recycling infrastructure, new market incentives, and more advanced sorting technologies, more and more construction waste has been diverted in recent years. Nevertheless, most diverted materials are valuable, easily resold materials such as metals, or high-volume waste, such as structural waste.

Both implementation and planning are essential for construction waste reduction. This credit encourages projects to implement the plan generated in MRP: Construction and Demolition Waste Management Planning. It sets thresholds for both a minimum number of material streams and an overall diversion percentage, and also encourages project teams to divert a greater diversity and quantity of materials into many material streams. The credit also offers an alternate to diversion via an option to reward waste reduction.

Refer to MRP: Construction and Demolition Waste Management Planning for more information.

Synergies
- MRP: Construction and Demolition Waste Management Planning

Important Notes for MR Category
1. Provide an easily accessible designated area for separation, collection, storage, and recycling of non-hazardous materials, like **p**aper, corrugated **c**ardboard, **m**etals, **g**lass, and **p**lastics for the entire building.

   Mnemonic: **P**eople **C**an **M**ake **G**reen **P**romises (See **bold** and underlined letters at the last part of the sentence listing the recycled materials.

   Note: These five materials required for recycling (as listed above) make up 59% of the total municipal solid waste stream. Food scraps (12%) and yard trimmings (13%) make up for 25% of the total. (The GBCI encourages you to compost these types of waste on-site if possible.) The remaining waste is wood (6%); textiles, leather, and rubber (7%); and other (3%).

2. Encourage the **3 R's**: reduce, reuse, and recycle.
3. Rapidly renewable materials include wool, bamboo, cotton, wheatboard, cork, strawboard, cotton insulation, agrifiber, linoleum, etc.

Chapter 11
Indoor Environmental Quality (EQ)

Overall Purpose
The EQ category encourages project teams to improve indoor air quality and visual, thermal, and acoustic comfort. LEED certified buildings protect the comfort and health of building occupants via good indoor environmental quality (EQ). Good EQ also improves the building's value, decreases absenteeism, enhances productivity, and reduces liability for building owners and designers. The EQ category covers many environmental design factors and strategies, such as control over one's surroundings, lighting quality, air quality, and acoustic design. These factors influence the way people live, learn, and work.

Researchers have not fully understood the complex relationship between the health and comfort of building occupants and the indoor environment. It is hard to measure and quantify the direct impact of a building on its occupants because of many variables including the building's site, design, and construction, occupant activities, and local customs and expectations. For this reason, the EQ section uses the performance-oriented credit requirements to balance the need for prescriptive measures. For instance, a prerequisite covers source control first, and a later credit then measures the actual outcome of those strategies with an indoor air quality assessment.

The EQ category takes advantage of both conventional methods, such as thermal and ventilation control, and emerging design strategies, including requirements for lighting quality (Interior Lighting credit), source control and monitoring for user-determined contaminants (Enhanced Indoor Air Quality Strategies credit), an emissions-based, holistic approach (Low-Emitting Materials credit), and advanced lighting metrics (Daylight credit). All projects using a BD+C rating system now have a new credit covering acoustics.

Cross-Cutting Issues
  Floor Area Calculations and Floor Plans
  The percentage of floor area meeting the credit requirements determines the compliance of many of the credits in the EQ category. Overall, space categorization and floor areas ought to be consistent across EQ credits. The project teams should explain and highlight any discrepancies or excluded spaces in floor area values in the documentation. See Space Categorization for more information on which floor area should be included.

  Space Categorization
  The EQ category concentrates on the interaction between the indoor spaces and the occupants of the building. Therefore, understanding which spaces are used by the building users and what activities they carry out in each space is critical. The credit requirements may or may not apply based on the space categorization (Table 11.1).

  Occupied Versus Unoccupied Space
  A space in a building is either occupied or unoccupied. Occupied spaces are for human activities. Unoccupied spaces are inactive areas for other purposes and occupied only occasionally and briefly. Typical unoccupied spaces include the following:
- electrical and mechanical rooms
- dedicated emergency exit corridor or egress stairway

- closets in a home (a walk-in closet is occupied space)
- data center floor area, including a raised floor area
- inactive storage area in a distribution center or warehouse

For areas with equipment retrieval, the space is unoccupied only if the retrieval is occasional.

Space Category	Prerequisite or Credit
Occupied space	• Minimum Indoor Air Quality Performance, ventilation rate procedure and natural ventilation procedure • Minimum Indoor Air Quality Performance, monitoring requirements • Enhanced Indoor Air Quality Strategies, Options 1C, 1D, 1E, 2B, 2E • Indoor Air Quality Assessment, Option 2, Air Testing (sampling must be representative of all occupied spaces) • Thermal Comfort (New Construction, Schools, Retail, Hospitality), design requirements • Acoustic Performance (New Construction, Data Centers, Warehouses and Distribution Centers, Hospitality)
Regularly occupied space	• Thermal Comfort, design requirements (Data Centers) • Interior Lighting, Option 2, strategies A, D, E, G, H • Daylight • Quality Views
Individual occupant space	• Thermal Comfort, control requirements • Interior Lighting, Option 1
Shared multi-occupant space	• Thermal Comfort, control requirements • Interior Lighting, Option 1
Densely occupied space	• Enhanced Indoor Air Quality Strategies, Option 2 C

Table 10.1   Space types in EQ credits

Regularly Versus Non-Regularly Occupied Spaces
Occupied spaces are divided into regularly occupied or non-regularly occupied categories according to the duration of the occupancy. **Regularly occupied spaces** are enclosed areas where building users typically spend at least one hour of continuous occupancy per person per day on average; the building users can be standing or seated as they study, work, or carry out other activities. For spaces not used daily, the classification ought to be according to the time a typical building user spends there when it is in use. For instance, a computer workstation can be vacant most the month, but when it is used, an occupant spends one to five hours in the space. It should be considered regularly occupied because that length of stay is sufficient to affect the occupant's health, and she or he would expect to have thermal control and comfort there.

Occupied spaces other than regularly occupied are **non-regularly occupied**; these are spaces that people use for less than one hour per person per day on average or simply pass through.

The following are examples of non-regularly occupied spaces:

• break room	• lobby (except hotel lobby)*
• circulation space	• locker room
• copy room	• residential bathroom
• corridor	• residential laundry area
• fire station apparatus bay	• residential walk-in closet
• hospital linen area	• restroom
• hospital medical record area	• retail fitting area
• hospital patient room bathroom	• retail stock room
• hospital short-term charting space	• shooting range
• hospital prep and cleanup area in surgical suite	• stairway
• interrogation room	

* Hotel lobbies are considered regularly occupied because people often work and spend more time there than in an office building lobby.

Occupied Space Subcategories

Occupied spaces, or portions of an occupied space, can also be divided into individual or shared multi-occupant per the number of building users and their activities. An **individual occupant space** is a place for someone to carry out distinctive tasks. A **shared multi-occupant space** is a space for congregation or for building users to perform collaborative or overlapping tasks. Non-regularly occupied spaces that are not used for collaborative or distinct tasks are *neither* shared multi-occupant *nor* individual occupant spaces.

Occupied spaces can also be divided into **densely** or **non-densely occupied** per the concentration of occupants in the space. A **densely occupied space** has no more than 40 square feet (3.7 square meters) per person, or 25 people or more per 1,000 square feet (93 square meters). A **non-densely occupied space** has more than 40 square feet (3.7 square meters) per person.

Rating System	Space Type	Prerequisite or Credit
Schools	classroom and core learning spaces	• Minimum Acoustic Performance • Acoustic Performance (Schools)
Hospitality	guest rooms	• Interior Lighting* • Thermal Comfort, control requirements*
Healthcare	patient rooms	• Thermal Comfort, control requirements • Interior Lighting, Option 2, Lighting Quality
	staff areas	• Interior Lighting, Option 2, Lighting Quality
	perimeter area	• Daylight • Quality Views
	inpatient units	• Quality Views
Warehouses & Distribution Centers	office areas	• Thermal Comfort, control requirements
	areas of bulk storage, sorting, and distribution	• Quality Views
Retail	office and administrative areas	• Thermal Comfort, control requirements • Interior Lighting, Option 2, Lighting Quality
	sales areas	• Interior Lighting, Option 2, Lighting Quality

Table 2. Rating system–specific space classifications

Space classifications will not affect the following credits:
- Environmental Tobacco Smoke Control
- Enhanced Indoor Air Quality Strategies, Option 1A, 1B, 2A, 2D (There are no specific spaces; applicable spaces are determined by the project team.)
- Low-Emitting Materials
- Construction Indoor Air Quality Management Plan
- Indoor Air Quality Assessment, Option 1, Flush-Out (The floor area from all spaces must be included in calculation for total air volume; the flush-out must be demonstrated at the system level.)
- Interior Lighting, Option 2, strategies B, C, and F
- Acoustic Performance (Healthcare)

The average American spends about 90% of his time indoors, so indoor environmental quality is very important for quality of life, well-being, and productivity.

EQ credits cover the following important topics:
1) Indoor environment quality (achieved through properly designed and installed systems)
2) Contaminants (reduce, manage, and eliminate)
3) Minimum IAQ performance
4) Environmental tobacco smoke (ETS) control
5) Outdoor air delivery monitoring
6) Increased ventilation
7) Construction IAQ management plan
8) Low-emitting materials
9) Indoor chemical and pollutant source control
10) Systems control
11) Thermal comfort
12) Daylight and views

Mnemonic
I Called Mike Evans.
Oh! Ian Catches LIST D (See underlined letters.)

Core Concepts
1) Improve indoor air quality
   - Improve building ventilation
   - Choose proper materials
   - Reduce, manage, and eliminate contaminants
   - Advocate green construction practices and green building operation
   - **Active (Mechanical)** Ventilation
   - **Passive (Natural)** Ventilation

2) Improve indoor environmental quality
   - Thermal comfort control
   - Daylight and views
   - Considering acoustics

Recognition, Regulation, and Incentives
Guidance is available from private and public sector organizations such as EPA, Department of Labor or ASHRAE

Overall Strategies and Technologies
1) Choose low-emitting materials, interior finishes, furniture, etc.
2) Reduce, manage, and eliminate contaminants like certain cleaning products, tobacco smoke, and radon.
3) Advocate green construction practices like **Best Management Practices (BMPs)**, outdoor air introduction, green cleaning, and proper handling of exhaust systems.
4) Allow thermal comfort control such as user control and feedback or operations and maintenance management.
5) Provide daylight/views such as north-facing skylight, interior light (reflecting) shelf, interior and exterior permanent shading devices, automatic photocell-based control, and high performance glazing. Try to maximize daylight for interior spaces while avoiding high-contrast conditions.
6) Use light fixtures with sensors and dimming controls.

EQP: Minimum Indoor Air Quality Performance

Mandatory for
    New Construction (0 points)
    Core and Shell (0 points)
    Schools (0 points)
    Retail (0 points)
    Data Centers (0 points)
    Warehouses and Distribution Centers (0 points)
    Hospitality (0 points)
    Healthcare (0 points)

Purpose

Establishes minimum standards for indoor air quality (IAQ), and contributes to the well-being and comfort of building occupants.

Prerequisite Concept

Ventilation dilutes pollutants created by occupants and other sources, and improves occupant well-being and comfort. There are strong associations between ventilation rates based on a multidisciplinary scientific review of the current knowledge and occupant health. Supplying at least some outdoor air, removing contaminants from outdoor air, and controlling pollutant sources, are some of the factors for maintaining good indoor air quality (IAQ). This prerequisite refers to the standards that outline well-tested methods for determining the amount of outdoor air needed for each type of space. The USGBC selects these standards because they balance the needs for maintaining energy efficiency and providing fresh air.

Different kinds of buildings, project type, equipment, activities, and occupants require different IAQ. For instance, residential projects have to comply with additional prescriptive requirements such as radon and combustion byproducts; and health care facilities have more rigorous space pressurization and ventilation criteria to prevent cross-contamination.

Intelligent ventilation design is only the beginning. This prerequisite requires monitoring to help maintain IAQ in all stages of a building's operation. As soon as proper ventilation stops, indoor air pollutants start to build up, even though building users may not notice decreases in outdoor airflow or exhaust airflow. Intelligent ventilation design and frequent monitoring helps to assure building user well-being and comfort.

Synergies
- EAP: Minimum Energy Performance
- EQC: Enhanced Indoor Air Quality Strategies
- EQC: Indoor Air Quality Assessment

EQP: Environmental Tobacco Smoke Control

Mandatory for
    New Construction (0 points)
    Core and Shell (0 points)
    Schools (0 points)
    Retail (0 points)
    Data Centers (0 points)
    Warehouses and Distribution Centers (0 points)
    Hospitality (0 points)
    Healthcare (0 point)

Purpose
Minimizes or protects ventilation air distribution systems, indoor surfaces, and building occupants from exposure to environmental tobacco smoke.

Prerequisite Concept
Every year over five million people die from tobacco use worldwide. Exposure to **environmental tobacco smoke (ETS)**, or secondhand smoke, also kills nonsmokers. In the United States, almost 50% of all nonsmokers were exposed to ETS regularly in 2006. Secondhand smoke increases nonsmokers' risk of developing many serious health problems such as heart disease and lung cancer.

Banning indoor smoking is the only method to completely eradicate the health problems related to ETS. Therefore, LEED-certified buildings do not allow designated indoor smoking rooms. The prerequisite also bans smoking in outdoor business areas such as a courtyard or public sidewalk for seating or kiosks. This is because a business still has control over the smoking policy in these areas, even if they may not be inside the property boundary line.

Prohibiting ETS in the building interior benefits human health, and improves the longevity of furniture, furnishings, air distribution systems, and building surfaces.

Synergies
    None

*EQP: Minimum Acoustic Performance

Applicable to
   Schools

Note: Detailed discussions have been omitted since this prerequisite is for Schools only, and is unlikely to be tested on the LEED Green Associate Exam.

EQC: Enhanced Indoor Air Quality Strategies

Applicable to
New Construction (1-2 points)
Core and Shell (1-2 points)
Schools (1-2 points)
Retail (1-2 points)
Data Centers (1-2 points)
Warehouses and Distribution Centers (1-2 points)
Hospitality (1-2 points)
Healthcare (1-2 points)

Purpose
Improves indoor air quality which results in occupant well-being, comfort, and productivity.

Credit Concept
Occupants bring particulates and pollutants indoors. They also come in through building openings or ventilation system intakes, and are generated by activities within the building. Designing for effective **indoor air quality (IAQ)** can improve indoor environment, and building occupants comfort and human health.

This credit expands the outdoor air requirements of EQ Prerequisite Minimum Indoor Air Quality Performance, and identifies additional IAQ strategies. Design strategies include the monitoring strategies for ventilation systems, increasing ventilation, using enhanced filtration media, and installing entryway systems to prevent occupants from bringing contaminants inside the building. The USGBC encourages a combination of multiple strategies.

Synergies
- EAP: Minimum Energy Performance
- EAC: Energy Performance
- EQP: Indoor Air Quality Performance

EQC: Low-Emitting Materials

Applicable to
- New Construction (1-3 points)
- Core and Shell (1-3 points)
- Schools (1-3 points)
- Retail (1-3 points)
- Data Centers (1-3 points)
- Warehouses and Distribution Centers (1-3 points)
- Hospitality (1-3 points)
- Healthcare (1-3 points)

Purpose

Lessens the concentration of chemical pollutants that may harm the environment, air quality, human productivity, and human health.

Credit Concept

Numerous kinds of chemicals are present in all places. Many materials release chemicals known as **volatile organic compounds (VOCs)** into the air. These materials are from many sources: human made, natural, plant-based, and from people and animals. Extended exposure to high intensity of some VOCs has been linked to cancer, chronic obstructive pulmonary disease, asthma, and many other chronic health problems. Short-term exposure to VOCs can trigger throat, nose, and eye irritation as well as other severe reactions.

The natural environment has some VOCs; nevertheless, indoor environments may have numerous sources of VOCs, reduced air ventilation, and can have higher concentrations of VOCs. It is impossible to totally eradicate exposure to all VOCs, however, specifying nonemitting products and low-emitting will considerably lessen the quantity and strength of indoor VOC exposure.

This credit encourages the selection of products that are inherently nonemitting or meet the compliance criteria established by recognized standards. For example, all paints and coatings wet-applied on site must meet the applicable VOC limits of the **South Coast Air Quality Management District (SCAQMD)** Rule 1113, effective June 3, 2011, or the **California Air Resources Board (CARB)** 2007, Suggested Control Measure (SCM) for Architectural Coatings. Preferably, all interior building materials should be in compliance, including all interior finishes, thermal and acoustic insulation, as well as furniture and furnishings. It uses a holistic systems approach, recognizes compliance of product assemblies even if some elements do not comply, and rewards teams for partial compliance.

This credit tackles each layer of the interior finish (wall, flooring, and ceiling). It conservatively protects occupants if project teams test the emissions from layers not directly exposed to air individually.

Compared with VOC content limits, air concentration measurements from chamber testing predict emissions over time better. However, chamber emissions testing cannot evaluate emissions during installation, is less widely used for wet-applied products, and is usually more expensive. The credit still limits VOCs for on-site wet-applied products in order to protect installers or those exposed to them at the time of application, and to avoid environmental damage such as smog formulation.

Synergies
- MRC: Furniture and Medical Furnishings (Healthcare only)
- EQC: Indoor Air Quality Assessment

EQC: Construction Indoor Air Quality Management Plan

Applicable to
- New Construction (1 point)
- Core and Shell (1 point)
- Schools (1 point)
- Retail (1 point)
- Data Centers (1 point)
- Warehouses and Distribution Centers (1 point)
- Hospitality (1 point)
- Healthcare (1 point)

Purpose
Lessens indoor air quality problems related to renovation and construction; and protects the health of building occupants and construction workers.

Credit Concept
Construction activities generate toxic substances, dust, or other contaminants, which adversely affect indoor air quality (IAQ), and cause long-term health problems for building occupants and construction workers. IAQ best practices during construction can protect building occupants from airborne pollutants, limit workers from exposure to dust and toxins during construction, and make HVAC equipment and building materials perform better and last longer.

One of the requirements for this credit is to develop and implement a construction IAQ management plan per the **Sheet Metal and Air Conditioning National Contractors' Association (SMACNA)** IAQ guidelines. The SMACNA standard lists main sources of construction-related indoor air pollution and clarifies best practices for controlling them.

With SMACNA IAQ strategies, buildings will capture airborne contaminants like dust, prevent mold and other harm to building materials, and keep toxic substances and pollutants out of building systems. Furthermore, projects have to make sure that all permanent air handlers operated during construction comply with filtration criteria, ban smoking during construction near entrances and inside the building, and protect absorptive material from moisture damage.

As a consequence of the unique needs of patients, health care facilities have extra criteria to address infection control, vibration, and noise that are above and beyond the basic SMACNA guidelines.

If you use a permanently installed air handler during construction, use filtration media with a Minimum Efficiency Reporting Value (MERV) of **8** at each return air grille per ASHRAE 52.2–2007, with errata (or equivalent filtration media class of F5 or higher, as defined by CEN Standard EN 779–2002, Particulate Air Filters for General Ventilation, Determination of the Filtration Performance). Change all filtration media immediately before occupancy.

Synergies
- EQC: Enhanced Indoor Air Quality Strategies
- EQC: Low-Emitting Materials
- EQC: Indoor Air Quality Assessment

EQC: Indoor Air Quality Assessment

Applicable to
　　New Construction (1-2 points)
　　Schools (1-2 points)
　　Retail (1-2 points)
　　Data Centers (1-2 points)
　　Warehouses and Distribution Centers (1-2 points)
　　Hospitality (1-2 points)
　　Healthcare (1-2 points)

Purpose
　　Sets up better IAQ in the building after construction and during occupancy.

Credit Concept

Many construction materials include substances that are harmful to human health; and construction activity can bring pollutants into the indoor environment. **Volatile organic compounds (VOCs)** and formaldehyde from construction materials are two of the harmful substances. Ozone, dust, and fine particles generated by unfiltered outdoor air, diesel engines, or construction activity can also be detrimental. Decreasing indoor air pollutants reduces absenteeism, improves productivity, has important health benefits, and usually improves building users' comfort.

The best method to ascertain that source control strategies have been properly and effectively accomplished is to test **airborne pollutant levels**. For VOCs, this credit uses the California Department of Public Health Standard Method v1.1, which has meticulous, well-established testing procedures and science-based best practices, and is widely acknowledged by the industry.

A building flush-out is an alternative to testing indoor air quality, and is an effective way to diffuse pollutants left behind from construction such as off-gassed compounds. A typical mechanical ventilation system is used as a base to set the threshold for duration of the building flush-out. Its typical supply airflow rate is 0.7 cubic feet per minute per square foot. For this reason, if a system operates at 100% outdoor air continuously for two weeks, the cubic feet of outdoor air per square foot of floor area is as follows:

14,112 cu ft of outdoor air / ft^2 of floor area =
{0.7 × (cfm/ ft^2) x 14 days × (24 hours/day) x (60 mins/hr)}

In SI units,
4,294,080 lps of outdoor air / sq meter of floor area
= {3.55 × (lps/m^2) × 14 days × (24 hours/day) x (60 mins/hr) × (60 sec/ min)}

This equation shows that two weeks of flush-out is sufficient time for removing pollutants from the construction phase.

Synergies
- EQP: Minimum Indoor Air Quality Performance
- EQC: Enhanced Indoor Air Quality Strategies
- EQC: Low-Emitting Materials
- EQC: Construction Indoor Air Quality Management

EQC: Thermal Comfort

Applicable to
- New Construction (1 point)
- Schools (1 point)
- Retail (1 point)
- Data Centers (1 point)
- Warehouses and Distribution Centers (1 point)
- Hospitality (1 point)
- Healthcare (1 point)

Purpose
Provides good thermal comfort, and improves occupant comfort, well-being, and productivity.

Credit Concept

Field and laboratory research has shown how the thermal conditions of buildings directly affect people's performance and satisfaction. While frequently correlated with air temperature only, thermal comfort is a complex amalgam of **six primary factors** including **surface temperature, air temperature, air movement, humidity, metabolic rate, and clothing**. All these factors are affected by building design and operation. Considering all six at the same time is an effective thermal comfort strategy. This requires close collaboration between the architect, engineer, and owner, which is essential to achieving this credit.

Changing one or more of the six comfort factors can reduce energy use and greatly improve building user experience of the thermal environment at the same time. Communicating effectively with the owner, the project team can coordinate design with operational policies and maximize comfort. For instance, a flexible dress code permits employees to wear appropriate clothing per the season, and allows design air temperatures to be higher in summer and lower in winter without affecting occupant comfort.

Allowing occupants to adjust their thermal environment via thermal controls will make them feel more comfortable regardless of the conditioning strategy, and will generate additional productivity and satisfaction. IAQ surveys conducted by the Center for the Built Environment have demonstrated major increases in satisfaction among occupants who have individual control of an operable window or a thermostat. Similarly, research administered by the International Centre for Indoor Environment and Energy indicates allowing occupants +/–5°F (3°C) of local temperature control can give rise to productivity gains of 2.7% to 7%.

Predicted percentage of dissatisfied (PPD) and **predicted mean vote (PMV)** are two indices used as the referenced standards for this credit. The PMV was created by placing people in climate chambers and asking them to use a seven-point thermal sensation scale to rate their level of comfort, from -3 (too cold) to +3 (too hot), with 0 representing neutral. The researchers then determined the PPD index. PPD predicts the percentage of people who will probably be dissatisfied with a given thermal condition.

This credit's referenced standards also use **field-based research** as the basis of the adaptive model, which links outdoor climatological or meteorological parameters to acceptable temperature ranges or indoor design temperatures.

Synergies
- EAP: Minimum Energy Performance
- EQP: Minimum Indoor Air Quality Performance
- EQC: Enhanced Indoor Air Quality Strategies
- EQC: Interior Lighting

EQC: Interior Lighting

Applicable to
New Construction (1–2 points)
Schools (1–2 points)
Retail (2 points)
Data Centers (1–2 points)
Warehouses and Distribution Centers (1–2 points)
Hospitality (1–2 points)
Healthcare (1 point)

Purpose
Provides high-quality lighting, and improves occupant comfort, well-being, and productivity.

Credit Concept
Research has shown that people are more productive and comfortable in a carefully illuminated environment where the individual or group has lighting controls to meet their needs. Also, high quality lighting helps reduce health problems, eliminates distractions, contributes to occupant well-being, supports communication and interaction, and creates a sense of place and visual interest. This credit encourages lighting quality that substantially improves occupant productivity and comfort.

The credit promotes lighting quality in many ways.
- **Strategy A** minimizes light fixture luminance to help reduce discomfort glare and disability. It uses 2,500 candela per square meter as the threshold, because glare becomes objectionable above that level according to research by the Light Right Consortium.
- **Strategy B** selects light sources with a color rendering index over 80 to help simulate natural light.
- **Strategy C** uses light sources with a long lamp life to help preserve the integrity of the lighting design over a long period; it also reduces resource and material inputs and lowers maintenance costs. A lamp life of 24,000 hours encourages the use of longer-life fluorescents.
- **Strategy D** designs spaces with less direct-only overhead lighting to help minimize glare, reduces the perceived brightness of the direct luminaires, and reduces contrast between ceiling and luminaires.
- **Strategies E and F** specify surfaces with high reflectance to use reflection to help make the space brighter, and minimize the difficulty of viewing light documents on dark surfaces. The specific surface reflectance values for floors, walls, and ceilings are above the standard industry assumptions of 20, 50, and 80, respectively, according to in the latest edition of the Illuminating Engineering Society (IES) Lighting Handbook.
- **Strategies G and H** design for an illuminance ratio less than 1:10 to minimize the amount of contrast that building users experience between their work surface or wall surfaces around them and the ceiling. The 1:10 illuminance ratio represents one log scale difference in lighting levels (illuminance is linear, but human eyes are logarithmic).

Synergies
- EQC: Thermal Comfort
- EAP: Fundamental Commissioning and Verification and EA Credit Enhanced Commissioning

EQC: **Daylight**

Applicable to
- New Construction (1–3 points)
- Core and Shell (1–3 points)
- Schools (1–3 points)
- Retail (1–3 points)
- Data Centers (1–3 points)
- Warehouses and Distribution Centers (1–3 points)
- Hospitality (1–3 points)
- Healthcare (1–2 points)

Purpose
Introduces daylight into the space, reduces the use of electrical lighting, and reinforces circadian rhythms by connecting building occupants with the outdoors.

Credit Concept
Increased exposure to daylight improves human health and behavior since it enhances our circadian rhythms. Sufficient daylight even increases sales in retail environments, increases productivity in the workplace, improves students' performance, helps healing in hospitals, and fights lethargy and depression. Daylight in buildings also conserves natural resources, uses less electric lighting energy, and reduces air pollution.

This credit has advanced greatly and now focuses on using **actual measurement** and **simulated daylight analysis** to approximate daylight levels and daylight quality. These techniques support the design process for optimizing daylight and more precisely predict daylight access. Past prescriptive methods for calculating daylight used **window design**, and less accurately considered such project-specific factors such as time of day and year, building orientation, the interaction with interior finishes, exterior conditions, and other performance variables. The new simulation requirements use performance values and global metrics for daylight created by daylighting professionals. Other internationally recognized standard-setting organizations are using the credit's performance goals, metric conversions, and language to create consistency in the building and daylighting professions.

Projects have three options for compliance. The options that demand more **detailed design analysis and input** or that establish **actual performance** earn a correspondingly higher number of points. The best optio to achieve compliance is to use a good **computer simulation** to inform the design phase and help build a more effective daylight project. In addition to integrating daylight concerns into the design process, project teams should also consider such factors as visual quality, glare control, heat gain and loss, and variations in daylight availability.

Synergies
- EAP: Minimum Energy Performance and EA Credit Optimize Energy Performance
- EQC: Quality Views
- EQC: Interior Lighting

EQC: Quality Views

Applicable to
    New Construction (1 point)
    Core and Shell (1 point)
    Schools (1 point)
    Retail (1 point)
    Data Centers (1 point)
    Warehouses and Distribution Centers (1 point)
    Hospitality (1 point)
    Healthcare (1-2 points)

Purpose
Provides quality views and gives building users a connection to the outdoor natural environment.

Credit Concept
Visually connecting with outdoor environments will give building occupants greater attentiveness, satisfaction, and productivity. Outside views with natural elements are more appealing and offer better visual relief. People sitting at computers often look at their screens for long periods without a break, and develop dry eyes or eyestrain. They find relief in attractive distant views. Similarly, providing patients with access and views to nature can reduce depression, stress, the use of pain medication, and shorten hospital stays.

Outside views also connect people with the changes in light from season to season, diurnal changes from light to dark, and other natural environmental cues. These are important for maintaining natural circadian rhythms. Interruption of these rhythms can cause mental disorders and other long-term health care problems.

To design for quality views, the project teams must consider site design and building orientation, interior layout, and facade, and use integrated design to identify potential compromises. For instance, glazing with tints, colors, patterns, fibers or frits is often used to enhance aesthetics, provide privacy, and reduce solar heat gain and glare; on the other hand, they can completely block the views to the outdoors.

The USGBC has incorporated four former exemplary performance paths as options for credit compliance, which expand the credit to consider the quality of views for building occupants. Glazing frit, color, and patterns have been restricted to maintain quality views. In addition, the type of objects in the view, such as sky, vegetation, busy streets, and brick walls, are now an important factor. Even though the standard has been raised, project teams have flexibility to choose from the four credit paths to comply. Also, up to 30% of the required area with access to quality views can now be into atriums, a change based on industry recognition that atriums can increase views and daylight for interior spaces, and reduce electrical lighting.

Synergies
- EQC: Daylight
- EAP: Minimum Energy Performance

EQC: Acoustic Performance

Applicable to
- New Construction (1 point)
- Schools (1 point)
- Data Centers (1 point)
- Warehouses and Distribution Centers (1 point)
- Hospitality (1 point)
- Healthcare (1-2 points)

Purpose
Creates effective acoustic design, and provides classrooms and workspaces that promote occupant communication, productivity, and well-being.

Credit Concept
As an indoor environmental quality imperative to complement other LEED practices, this credit encourages project teams to use best practices in acoustic design.

According to research by the Center for the Built Environment (CBE) of 34,000 building inhabitants interviewed, LEED buildings perform better than conventional buildings in all areas of indoor environmental quality except acoustics. There are trade-offs between acoustic performance and other green building practices, such as efficient lighting strategies, highly efficient HVAC systems, and open floor planning.

Because acoustics significantly affect healing and learning environments, LEED 2009 covered it in the Healthcare and Schools rating systems. LEED v4 includes an acoustics credit for all new construction projects, encouraging project teams to balance acoustical design strategies with considerations for thermal comfort, daylighting, and other performance areas for planning systems and indoor spaces. For all LEED rating systems, well designed acoustics can improve the well-being of workers, facilitate communication, increase productivity, or aid in noise control and speech privacy. All of these enhance the environmental quality of the space.

Synergies
- For Schools, EQP: Minimum Acoustic Performance

Chapter 12
Innovation (IN)

Overall Purpose

Sustainable design measures and strategies are continuously improving and evolving. Contemporary scientific research affects building design strategies, and new technologies constantly emerge in the marketplace. This LEED credit category is to acknowledge projects for innovation in sustainable building practices, strategies, and features.

Sometimes, a strategy can help a building perform much better than the existing LEED credit requirements. Other strategies should be considered for their sustainable benefits, but they may not be covered by any LEED prerequisite or credit. This category also addresses the role of a LEED Accredited Professional as part of a cohesive team in implementing the cohesive process.

Innovation Credit (INC): **Innovation**

Applicable to
    New Construction (1-5 points)
    Core and Shell (1-5 points)
    Schools (1-5 points)
    Retail (1-5 points)
    Data Centers (1-5 points)
    Warehouses and Distribution Centers (1-5 points)
    Hospitality (1-5 points)
    Healthcare (1-5 points)

Purpose
Encourages projects to achieve innovative or exceptional performance.

Credit Concept
Sustainable design originates from innovative thinking and strategies. This credit and other institutional measures reward innovative thinking, and benefit our environment. Acknowledgment of exceptional performance will stimulate more innovation.

When project teams go above and beyond LEED requirements and innovate, they achieve measurable environmental benefits exceeding LEED rating system criteria, and have the chance to help the development of future LEED credits and explore the latest pilot credits. When they can show that the project exceeds the standard level of performance related to one or more LEED credits, their innovations can be implemented by other teams in the future.

For this credit, project teams can attempt the following points:
- Innovation (up to 4 points)
- Pilot credits (up to 4 points)
- Exemplary Performance (up to 2 points)

Synergies
    None

INC: LEED Accredited Professional

Applicable to
- New Construction (1 point)
- Core and Shell (1 point)
- Schools (1 point)
- Retail (1 point)
- Data Centers (1 point)
- Warehouses and Distribution Centers (1 point)
- Hospitality (1 point)
- Healthcare (1 point)

Purpose
Encourages the team assimilation to meet a LEED project requirements, and streamlines the application and certification process.

Credit Concept
A minimum of one *principal* participant of the project team must be a LEED Accredited Professional (AP) with a specialty suitable for the project.

A LEED AP with specialty is a valuable resource in the LEED certification process. She or he helps the project team members understand the LEED application process, the rating system, and the importance of synergies (i.e., the interactions among the prerequisites and credits).

Synergies
None

Chapter 13
Regional Priority (RP)

Overall Purpose

Since every locale has some unique environmental issues, volunteers from the LEED International Roundtable and USGBC chapters have pinpointed specific environmental priorities within their areas and the credits that address them. These RP credits encourage project teams to concentrate on their local environmental priorities.

USGBC set up a procedure that pinpointed six RP credits for every rating system and every location within country boundaries or chapter. USGBC asked participants to decide which environmental issues were most significant in their country or chapter area. The issues could be man-made, such as polluted watersheds, or naturally occurring, such as water shortages, and could reflect environmental assets, such as abundant sunlight, or environmental concerns, such as water shortages. The zones, or areas, were identified by a mixture of priority issues, such as an urban area with an intact watershed versus an urban area with an impaired watershed. USGBC then asked the participants to prioritize credits to tackle the critical issues of certain locations. Because each LEED project type, such as a data center, may be related to distinctive environmental impacts, each rating system has its own RP credits.

The ultimate goal of RP credits is to improve the ability of LEED project teams to tackle essential environmental issues around the world and across the country.

Regional Priority Credit (RPC): **Regional Priority**

Applicable to
- New Construction (1-4 points)
- Core and Shell (1-4 points)
- Schools (1-4 points)
- Retail (1-4 points)
- Data Centers (1-4 points)
- Warehouses and Distribution Centers (1-4 points)
- Hospitality (1-4 points)
- Healthcare (1-4 points)

Purpose
Offers an incentive to achieve credits that cover geographically specific public health, social equity, and environmental priorities.

Credit Concept
Project teams design, build, and operate LEED buildings in many different circumstances. Local regulations, population density, and climate can vary drastically from one site to another, making some environmental issues more significant than others. Rainwater management in wet climates versus water conservation in arid climates is one example.

If project teams identify their location's priority environmental issues and tackle them in design, construction, and operation, LEED projects can be more transformative. LEED uses RP credits to promote a focus on regional issues. The current RP LEED credits are those that USGBC volunteers have decided to be especially important in a certain area. For every location in the United States, volunteers prioritize six credits. The ultimate goal is to encourage project teams to earn the credits that tackle an area's priority issues.

See the RP credit database at the following link:
http://www.usgbc.org

You can obtain one to four credits out of the six potential RP points. You can choose which four credits you want to pursue for your project.

The USGBC has prioritized the projects located in the United States, the US Virgin Islands, Guam, and Puerto Rico. Project teams for other international projects can check the database above for eligible RP points.

Synergies
None

Chapter 14
LEED Green Associate Exam (LEED-GA) Sample Questions, Answers and Exam Registration

I. LEED Green Associate Exam sample questions

We have provided twenty sample questions in this book to give you an idea what the USGBC is looking for on the LEED Green Associate Exam, and how the questions will be asked. These sample questions are quite easy. If you can answer 80% of the sample questions correctly, you are ready to take more mock exams. (The 80% passing score is based on feedback from previous readers.) We have published a separate book "LEED Green Associate Mock Exams" which has two hundred questions divided into two mock exams. You really need to review the study materials several times and MEMORIZE the important information before you attempt these mock exams. Just like on the real exam, sometimes a question may ask you to pick two or three correct answers out of four, or four correct answers out of five (some LEED exam questions have five choices). This means that if you do not know any one of the correct answers, you will probably get the overall answer wrong. You need to know the LEED system very well to get the correct answer.

There is one exception that I make which may be different from the real LEED Green Associate Exam: I broke the main LEED credit category into separate credits. This is because I believe most people will take the LEED AP+ or LEED AP with specialty exam at some point after they take the LEED Green Associate Exam.

By organizing this book by each specific LEED credit, I make it easier for you to support a LEED project team because people will talk about specific credits when they do real LEED projects.

Another benefit is that you only need to spend 50% more effort to prepare for Part Two of the LEED AP+ exam since you already know most of specific LEED credits. The basic and fundamental credit system is very similar for ALL LEED systems.

If I did not break the main LEED category into separate credits, you would have to double your effort to prepare for Part Two of the LEED AP+ exam by having to MEMORIZE all the information twice: once for each main LEED category and again per each specific LEED credit.

It is much easier to understand, digest, and memorize the information by specific LEED credit instead of by the main LEED category.

1. With regard to the Optimize Energy Performance credit, which of the following statements is correct?
    a. Compare your building performance to the baseline building performance.
    b. Compare your baseline building performance to ASHRAE Standard 90.1-2007 (with errata but without addenda).
    c. Compare your baseline building performance to ASHRAE Standard 90.1-2004.
    d. Compare your baseline building performance to ASHRAE Standard 90.1-2003.

2. Which of the following factors does not improve human comfort?
    a. air temperature
    b. ventilation
    c. radiation exchange
    d. all of the above

3. The standard used for Measurement and Verification is
    a. ASHRAE Standard 90.1-2007 (with errata but without addenda)
    b. the Department of Energy Verification Protocol
    c. a signed statement from the designer
    d. International Performance Measurement and Verification Protocol

4. Optimal IAQ performance can do which of the following?
    a. Create savings in electrical bills.
    b. Improve the productivity and health of building occupants.
    c. Cause higher operation cost.
    d. Create higher rents.

5. What is the best way to control Environmental Tobacco Smoke (ETS)?
    a. Ban smoking inside the building.
    b. Place exterior smoking areas at least 25 feet away from operable windows, entrances and air intakes.
    c. Give all interior spaces a negative pressure.
    d. both a and b

6. With regard to the Construction IAQ Management Plan, if your project uses permanently installed air handlers during construction, you should use
    a. filtration media with a minimum Efficiency Reporting Value (MERV) of 6 at each return air grille
    b. filtration media with a minimum Efficiency Reporting Value (MERV) of 7 at each return air grille
    c. filtration media with a minimum Efficiency Reporting Value (MERV) of 8 at each return air grille
    d. filtration media with a minimum Efficiency Reporting Value (MERV) of 9 at each return air grille

Chapter 14: Sample Questions, etc. • 155

7. For EQC: Enhanced Indoor Air Quality Strategies, which of the following are not design strategies? **(Choose two)**
   a. Increase ventilation.
   b. Submit drawings and cutsheets for the plumbing systems in chemical mixing areas.
   c. Prevent occupants from bringing contaminants inside the building by installation of entryway systems.
   d. Design the exterior sidewalk and pavement to drain away from the building at 2% minimum slope.

8. Placing a lighting control in a hallway does not help you with Controllability of Systems because
   a. You need to make sure the hallway does not have a dead end corridor that is over 20 feet in length.
   b. You also need to add temperature control.
   c. You also need to make sure the hallway has a view to the outside.
   d. none of the above.

9. Which of the following standard(s) is mentioned with regard to EQC: Construction Indoor Air Quality Management Plan?
   a. ASHRAE Standard 52.2-1999
   b. ASHRAE Standard 62.1-2007
   c. ASHRAE Standard 90.1-2007 (with errata but without addenda)
   d. SMACNA
   e. SCAQMD Rule 1168

10. Which of the following standard(s) is (are) mentioned in SSC: Light Pollution Reduction?
    a. ASHRAE/IESNA Standard 90.1-2007 (with errata but without addenda)
    b. IESNA RP-33
    c. International Dark Sky Association Outdoor Lighting Standard
    d. BUG
    e. both a and b

11. California Title 24-2013 is considered to be equal to ASHRAE/IESNA Standard 90.1-2010(with errata but without addenda) for the following LEED BD+C rating systems credit(s)
    a. EAP: Minimum Energy Performance
    b. EAC: Optimize Energy Performance
    c. EAC: Green Power and Carbon Offsets
    d. all of the above

12. Which of the following is not a consideration for MRC: Building Product Disclosure and Optimization—Sourcing of Raw Materials?
    a. the harvest cycle of the raw materials
    b. the Sustainable Agriculture Standard
    c. corporate sustainability reports (CSRs)
    d. USGBC-approved program

13. What does "Xeriscape" mean?
    a. Drip irrigation to save water.
    b. "Dry Landscape" design by using plants that use little or no water.
    c. Recycle existing plants on the project site.
    d. Reuse graywater for landscape irrigation.

14. Which of the following is a responsibility for the contractor to support the LEED documentation process?
    a. Document and provide calculations for waste diverted from landfill.
    b. Maintain a submittal log.
    c. Maintain a RFI log.
    d. Provide written documentation to justify a change order for rough grading.

15. Materials that qualify for MRC: Building Product Disclosure and Optimization—Sourcing of Raw Materials may also qualify for which of the following credit(s)?
    a. MRC: Building Life Cycle Impact Reduction
    b. MRC: Building Product Disclosure and Optimization—Environmental Product Declarations
    c. MRC: Building Product Disclosure and Optimization—Material Ingredients
    d. all of the above

16. Which of the following three statements is incorrect?
    a. For non-residential projects, water closet uses per day per FTE female is three.
    b. For residential projects, water closet uses per day per female is five.
    c. The flow rate for a conventional water closet is 1.8 gpf.
    d. The flow rate for a low-flow water closet is 1.1 gpf.

17. Which of the following is the standard for paints and coatings wet-applied on site? **(Choose two)**
    a. The California Air Resources Board (CARB) 2007, Suggested Control Measure (SCM) for Architectural Coatings
    b. The South Coast Air Quality Management District (SCAQMD) Rule 1113, effective June 3, 2011
    c. Green Seal Standard for Commercial Adhesives GS-36 requirements
    d. Green Label Plus
    e. Green Guard

18. Why should a developer locate a green building in a previously developed urban area?
    a. to be close to public transportation
    b. to use existing community services
    c. to be close to existing utilities
    d. a and b
    e. a, b, and c

19. Blackwater is water drained from a
    a. kitchen sink
    b. toilet
    c. both a and b
    d. none of the above

20. With regard to ozone depletion potential (ODP), the order from high ODP to low ODP is
    a. CFC> HFC>HCFC
    b. CFC>HCFC>HFC
    c. HCFC>HFC>CFC
    d. HCFC> CFC>HFC

II. Answers for the LEED Green Associate Exam sample questions

1. Answer: a. Compare your building performance to the baseline building performance.
   For Optimize Energy Performance credit, there are two ways to gain the credit, through a prescriptive approach or performance approach. Performance approach uses **whole building energy simulation** to compare your building performance to the baseline building and awards points based on percentage improvement in energy performance.

   See "EAC: Optimize Energy Performance" for more information.

   The following are distracters to confuse you:
   - Compare your baseline building performance to ASHRAE Standard 90.1-2007 (with errata but without addenda).
   - Compare your baseline building performance to ASHRAE Standard 90.1-2004.
   - Compare your baseline building performance to ASHRAE Standard 90.1-2003.

2. Answer: c. radiation exchange
   Pay attention to the word "not."

   While frequently correlated with air temperature only, thermal comfort is a complex amalgam of **six primary factors** including **surface temperature, air temperature, air movement, humidity, metabolic rate, and clothing**. See EQC: Thermal Comfort for more information.

   The following factors do improve human comfort, and therefore are the incorrect answers.
   - air temperature
   - ventilation (air movement)

3. Answer: d. International Performance Measurement and Verification Protocol

   This information is pretty obscure, and not covered by our previous discussion. You will probably run into a very small number of questions that you have no clue on the correct answer for in the real test. We intentionally include this hard question with obscure information at the beginning of the sample questions to test your ability to manage your time. You need to be able to manage your time, pick a guess answer for a very hard question, move on to the other questions, and come back to review it later if you have time.

   The following are distracters to confuse you:
   - ASHRAE Standard 90.1-2007 (with errata but without addenda)
   - the Department of Energy Verification Protocol
   - a signed statement from the designer

4. Answer: b. Improve the productivity and health of building occupants.

   Optimal Indoor Air Quality (IAQ) performance will not create savings in electrical bills, and does not necessary cause higher operation cost, or create higher rents.

5. Answer: d. both a and b
   See EQP: Environmental Tobacco Smoke Control

   Giving all interior spaces a negative pressure is not the best way to control Environmental Tobacco Smoke (ETS).

6. Answer: c. filtration media with a minimum Efficiency Reporting Value (MERV) of 8 at each return air grille

   See EQC: Construction Indoor Air Quality Management Plan.

7. Answer: b and d
   Pay attention to the word "not."

   For EQC: Enhanced Indoor Air Quality Strategies, both of the following are not design strategies, and therefore the correct answers:
   - Submit drawings and cutsheets for the plumbing systems in chemical mixing areas.
   - Design the exterior sidewalk and pavement to drain away from the building at 2% minimum slope.

   The following are correct design strategies but incorrect answers:
   - Increase ventilation.
   - Prevent occupants from bringing contaminants inside the building by installation of entryway systems.

8. Answer: d. none of the above
   See EQC: Thermal Comfort.

   Placing a lighting control in a hallway does not help you with Controllability of Systems because a hallway is a common transportation space, and does not give occupants a sense of control over their own space.

   The following are distracters to confuse you:
   - You need to make sure the hallway does not have a dead end corridor that is over 20 feet in length.
   - You also need to add temperature control.
   - You also need to make sure the hallway has a view to the outside.

9. Answer: d. SMACNA
   See EQC: Construction Indoor Air Quality Management Plan. One of the requirements for this credit is to develop and implement a construction IAQ management plan per the **Sheet Metal and Air Conditioning National Contractors' Association (SMACNA)** IAQ guidelines.

   The following are distracters to confuse you:
   - ASHRAE Standard 52.2-1999
   - ASHRAE Standard 62.1-2007
   - ASHRAE Standard 90.1-2007 (with errata but without addenda)
   - SCAQMD Rule 1168

10. Answer: d. BUG
    See SSC: Light Pollution Reduction. This credit offers two choices to give designers flexibility: **a new backlight, uplight, and glare (BUG) rating method;** and **a calculation method** (as in LEED 2009).

    The following are distracters to confuse you:
    - ASHRAE/IESNA Standard 90.1-2007 (with errata but without addenda)
    - IESNA RP-33
    - International Dark Sky Association Outdoor Lighting Standard ("Outdoor Lighting Standard" is an invented term that does not exist at all. The correct term is Illuminating Engineering Society and International Dark Sky Association (IES/IDA) Model Lighting Ordinance (MLO) User Guide and IES TM-15-11, Addendum A. For further info see ies.org.)

11. Answer: d. all of the above
    California Title 24-2013 is considered to be equal to ASHRAE/IESNA Standard 90.1-2010 (with errata but without addenda) for all the following LEED BD+C rating systems credit(s):
    - EAP: Minimum Energy Performance
    - EAC: Optimize Energy Performance
    - EAC: Green Power and Carbon Offsets

12. Answer: a. the harvest cycle of the raw materials
    Pay attention to the word "not."

    See MRC: Building Product Disclosure and Optimization—Sourcing of Raw Materials.

    This is a change from LEED 2009. Bio-based materials are *not* defined by the harvest cycle of the raw materials anymore; instead, products must meet the Sustainable Agriculture Standard to count toward this credit.

    The following are considerations for MRC: Building Product Disclosure and Optimization—Sourcing of Raw Materials, and therefore *not* the correct answers:
    - the Sustainable Agriculture Standard
    - corporate sustainability reports (CSRs)
    - USGBC-approved program

13. Answer: b. "Dry Landscape" design by using plants that use little or no water.

    "Xeriscape" is a very common term used in sustainable landscape practice, and you need to know about it.

    The following are distracters to confuse you:
    - Drip Irrigation to save water.
    - Recycle existing plants on the project site.
    - Reuse graywater for landscape irrigation.

Chapter 6 · 161

14. Answer: a. Document and provide calculations for waste diverted from landfill.
    The following are typical responsibilities for the contractor in a standard construction practice but are not a responsibility for the contractor to support the LEED documentation process.
    * Maintain a submittal log.
    * Maintain a RFI log.
    * Provide written documentation to justify a change order for rough grading.

15. Answer: d. all of the above
    See synergies for MRC: Building Product Disclosure and Optimization—Sourcing of Raw Materials.

    Materials that qualify for MRC: Building Product Disclosure and Optimization—Sourcing of Raw Materials may also qualify for all of the following credit(s):
    * MRC: Building Life Cycle Impact Reduction
    * MRC: Building Product Disclosure and Optimization—Environmental Product Declarations
    * MRC: Building Product Disclosure and Optimization—Material Ingredients

16. Answer: c
    See "Some use information for WE Category" at the end of WE Category.
    The flow rate for a conventional water closet is actually 1.6 gpf, instead of 1.8 gpf.

    The following three statements are correct:
    * For non-residential projects, water closet uses per day per FTE female is three.
    * For residential projects, water closet uses per day per female is five.
    * The flow rate for a low-flow water closet is 1.1 gpf.

17. Answer: a and b
    See EQC: Low-Emitting Materials.

    The following are standards for paints and coatings wet-applied on site:
    * The California Air Resources Board (CARB) 2007, Suggested Control Measure (SCM) for Architectural Coatings
    * The South Coast Air Quality Management District (SCAQMD) Rule 1113, effective June 3, 2011

    The following are distracters to confuse you:
    * **Green Seal Standard for Commercial Adhesives GS-36 requirements** (Green Seal certification is for identifying metal-free paints or an equivalent source of lead- and cadmium-free documentation)
    * **Green Label Plus** (This is a Carpet and Rug Institute (CRI) standard for carpet and adhesives.)
    * **Green Guard** ("GREENGUARD Certification is part of UL Environment, a business unit of UL (Underwriters Laboratories). GREENGUARD Certification helps manufacturers create and helps buyers identify interior products and materials that have low chemical emissions, improving the quality of the air in which the products are used." See http://www.greenguard.org for more information.)

18. Answer: e
All of the following are reasons why a developer should locate a green building in a previously developed urban area:
- to be close to public transportation
- to use existing community services
- to be close to existing utilities

19. Answer: c
Water drained from a kitchen sink or toilet is considered blackwater because it is contaminated by human waste or grease.

20. Answer: b
With regard to ozone depletion potential (ODP), the order from high ODP to low ODP is CFC>HCFC>HFC.

See the FREE PDF entitled "The Treatment by LEED® of the Environmental Impact of HVAC Refrigerants (LEED Technical and Scientific Advisory Committee, 2004)." This is one of the documents listed in the latest LEED Green Associate Exam Candidate Handbook. You should read all the documents listed by the handbook a few times and become familiar with them.

III. Where can I find the latest official sample questions for the LEED Green Associate Exam?

Answer: You can find them, as well as the exam content from the candidate handbook, at http://www.gbci.org/main-nav/professional-credentials/resources/candidate-handbooks.aspx.

IV. LEED Green Associate Exam registration

1. **How to register for the LEED Green Associate Exam?**
   Answer: Per the GBCI, you must create an Eligibility ID at www.GBCI.org. Select the "Schedule an Exam" menu to set up an exam time and date with Prometric. You can reschedule or cancel the LEED Green Associate Exam at www.prometric.com/gbci with your Prometric-issued confirmation number for the exam. You need to bring two forms of ID with you to the exam site. See www.prometric/gbci for a list of exam sites. Call 1-800-795-1747 (within the US) or 202-742-3792 (Outside of the US) or e-mail exam@gbci.org if you have any questions.

2. **Important Note:** You can download the "LEED Green Associate Candidate Handbook" from the GBCI website and get all the latest details and procedures. Ideally you should download it and read it carefully at least three weeks before your exam. See the link below:
   http://www.gbci.org/main-nav/professional-credentials/resources/candidate-handbooks.aspx

Chapter 15
Frequently Asked Questions (FAQ) and Other Useful Resources

The following are tips on how to pass the LEED exam on the first try and in one week. Also included are my responses to some readers' questions that may help you.

1. I found the reference guide way too tedious. Can I only read your book and just refer to the USGBC reference guide (if one is available for the exam I am taking) when needed?

Response: Yes, that is one way to study.

2. Is one week really enough for me to prepare for the exam while I am working?

Response: Yes, if you can put in 40 to 60 hours during the week, study hard and you can pass the exam. This exam is similar to a history or political science exam; you need to MEMORIZE the information. If you take too long, you will probably forget the information by the time you take the test.
In my book, I give you tips on how to MEMORIZE the information, and I have already highlighted/underlined the most important information that you definitely have to MEMORIZE to pass the exam. The intention of this book is to help you to pass the LEED exam with minimum time and effort. I want to make your life easier.

3. Will your book(s) be adequate for me to pass the exam?
Would you say that if I buy your LEED Exam Guide series books, I could pass the exam using no other study materials? The books sold on the USGBC website run in the hundreds of dollars, so I would be quite happy if I could buy your book and just use that.

Response: First of all, there are readers who have passed the LEED exam by reading *only* my books in the LEED Exam Guide series (www.GreenExamEducation.com or www.GeeForums.com). My goal is to write one book for each of the LEED exams, and make each of my books stand alone to prepare people for one specific LEED exam.

Secondly, people learn in many different ways. That is why I have added some new advice below for people who learn better by doing practice tests.

If you adhere to the following tips, you have a very good chance of passing the LEED exam (this is *not* a guarantee, *nobody* can guarantee you will pass):
a. Understand and *memorize* all of the information in my book, do *not* panic when you run into problems you are not familiar with, and use the guess strategy described in my book when taking the real exam.

You need to *understand* and *memorize* the information in the book and score almost a perfect score on the mock exam in this book. This book will give you the *bulk* of the most *current* information that you need for the specific LEED exam you are taking. You have to know the information in my book in order to pass the exam.

b. If you have not done any LEED projects before, I suggest you also go to the USGBC website and download the latest LEED credit templates for the LEED rating system related to the LEED exam you are taking. Read the templates and become familiar with them. This is important.
See the following link:
http://www.usgbc.org/leed#rating

c. If you want to be safe and take additional sample tests to find out if you are ready for the real exam, we have other books available on various LEED mock exams, including *LEED GA Mock Exams*. Check them out at the following link:
www.GreenExamEducation.com

The sample tests in *LEED GA Mock Exams* are very close to the real exam.

The LEED exam is NOT an easy exam, but anyone with a seventh grade education should be able to study and pass the LEED exam if he prepares correctly.

If you have extra time and money, the other books I would recommend are *LEED GA Mock Exams* and the USGBC reference guide pertaining to the LEED specialty you wish to pursue. I know some people who did not even read the reference guide from cover to cover when they took the exam. They just studied the information in my book, and only referred to the reference guide to look up a few things, and they passed on the first try. Some of my readers have even passed *without* looking at a USGBC reference guide *at all*.

4. I am preparing for the LEED exam. Do I need to read the 2" thick reference guide?

Response: See the answer above.

5. For LEED v4, will the total number of points be more than 110 in total if a project gets all of the extra credits and all of the standard credits?

Response: No. For LEED v4, there are 100 base points and 10 possible bonus points. There are many different ways to achieve bonus points (extra credits or exemplary performance), but you can have a maximum number of 6 Innovation (IN) bonus points and 4 Regional Priority (RP) bonus points. So, the maximum points for ANY project will be 110.

6. Are you writing new versions of books for the new LEED exams? What new books are you writing?

Response: Yes, I am working on other books in the LEED Exam Guide series. I will be writing one book for each of the LEED AP specialty exams. See GreenExamEducation.com or GeeForums.com for more information.

7. Important documents that you need to download for <u>free</u>, become familiar with, and <u>memorize</u>

Note: GBCI and USGBC changes the links to their documents every now and then, so, by the time you read this book, they may have changed some of those listed in this book. You can simply go to their main website, search for the document or subject by name, and should be able to find the most current link. You can use the same technique to search for documents by other organizations.

The main website for the GBCI is at the following link:
http://www.gbci.org/

The main website for the USGBC is at the following link:
http://www.usgbc.org/

a. Every LEED exam **always tests** Credit Interpretation Request (CIR). Download the guidelines for CIR customers, read, and memorize.
   See http://www.gbci.org/Certification/Resources/cirs.aspx.

b. Every LEED exam **always tests** project team coordination. Download *Sustainable Building Technical Manual: Part II*, by Anthony Bernheim and William Reed (1996), read, and memorize.
   See http://www.gbci.org/Files/References/Sustainable-Building-Technical-Manual-Part-II.pdf.

c. Project registration application and LEED certification process can be found at the following link:
   http://www.usgbc.org/leed/certification#tools

d. LEED Online can be found at the following link:
   http://www.usgbc.org/leed/certification#tools

8. Important documents that you need to download for free, and become familiar with:

a. *LEED for Operations and Maintenance Reference Guide-Introduction* (v4)
   http://www.usgbc.org/sites/all/assets/section/files/v4-guide-excerpts/Excerpt_v4_OM.pdf

b. *LEED for Operations and Maintenance Reference Guide-Glossary* (US Green Building Council, 2008)
   http://www.gbci.org/Files/References/LEED-for-Operations-and-Maintenance-Reference-Guide-Glossary.pdf

c. *LEED for Homes Rating System* (US Green Building Council, 2008)
   http://www.gbci.org/Files/References/LEED-for-Homes-Rating-System.pdf

   Pay special attention to the list of **abbreviations and acronyms** on pages 105–106 and a helpful **glossary of terms** on pages 107–114.

d. *Cost of Green Revisited,* by Davis Langdon (2007)
   http://www.gbci.org/Files/References/Cost-of-Green-Revisited.pdf

e. *The Treatment by LEED® of the Environmental Impact of HVAC Refrigerants* (LEED Technical and Scientific Advisory Committee, 2004)
   http://www.gbci.org/Files/References/The-Treatment-by-LEED-of-the-Environmental-Impact-of-HVAC-Refrigerants.pdf

f. *Guidance on Innovation and Design (ID) Credits* (US Green Building Council, 2004)
   http://www.gbci.org/Files/References/Guidance-on-Innovation-and-Design-Credits.pdf

9. Do I need to take many practice questions to prepare for a LEED exam?

Response: There is *no* absolutely correct answer to this question. People learn in many different ways. Personally, I am *not* crazy about doing many practice questions. Consider if you do 700 practice questions, not only must you read them all, and each question has at least 4 choices. This means there are a total of at least 2,800 choices, which is a great deal of reading. I have seen some third-party materials that have 1,200 practice questions. These materials will require even *more* time to go over the information.

I prefer to spend most of my time reading, digesting, really understanding the fundamental materials, and *memorizing* them naturally by rereading the text multiple times. The fundamental materials for *any* exam will *not* change, and the scope of the exam will *not* change for the same main version of the test (until the exam moves to the next advanced version). However, there are multiple ways to ask questions about this fundamental material.

If you have a limited amount of time for preparation, it is more efficient for you to focus on the fundamental materials and actually *master* the knowledge that GBCI wants you to learn. If you can do this, then no matter how GBCI changes the exam format or how GBCI asks the questions, you will do fine in the exam.

Strategy 101 for the LEED Green Associate Exam is that you must recognize that you have only a limited amount of time to prepare for the exam. Therefore, you must concentrate on the most important contents of the LEED Green Associate Exam.

The key to passing the LEED Green Associate Exam, or any other exam, is to know the scope of the exam, and not to read too many books. Select one or two helpful books and focus on them. You must understand the content and memorize it. For your convenience, I have underlined the fundamental information that I think is very important in this book. You definitely need to memorize all the information that I have underlined. You should try to understand the meaning first, and then memorize the content of the book by rereading it. This is a much better way than "mechanical" memory without understanding.

Most people fail the exam NOT because they are unable to answer the few "advanced" questions on the exam, but because they have read the information and can NOT recall it on the day of the exam. They spend too much time preparing for the exam, drag the preparation process on too long, seek too much information, go to too many websites, do too many practice questions and too many mock exams (one or two sets of mock exams are probably sufficient), and spread themselves too thin. They end up missing out on the most important information of the LEED exam, and they fail.

For me, memorization and understanding work hand-in-hand. Understanding always comes first. If you really understand something, then memorization is easy.

For example, I'll read a book's first chapter very slowly but make sure I *really* understand everything in it, no matter how long it takes. I do NOT care if others are faster readers than I. Then, I reread the first chapter again. This time, the reading is so much easier, and I can read it much faster. Then I try to recall the contents, focusing on substance, not the format or any particular order of things. This is a very good way for me to understand and digest the material, while *absorbing* and *memorizing* the content.

I then repeat the same procedure for each chapter, and then reread the book until I take the exam. This achieves two purposes:

a. I keep reinforcing the important materials that I already have memorized and fight against the human brain's natural tendency to forget things.

b. I also understand the content of the book much better by reading it multiple times.

If I were to attempt to memorize something without understanding it first, it would be very difficult for me to do so. Even if I were to memorize it, I would likely forget it quickly.

Appendixes

1. **Default occupancy factors**

Occupancy	Gross sf per occupant	
	Transient Occupant	**FTE**
Educational, Daycare	630	105
Educational, K–12	1,300	140
Educational, Postsecondary	2,100	150
Grocery store	550	115
Hotel	1,500	700
Laboratory or R&D	400	0
Office, Medical	225	330
Office, General	250	0
Retail, General	550	130
Retail or Service (auto, financial, etc.)	600	130
Restaurant	435	95
Warehouse, Distribution	2,500	0
Warehouse, Storage	20,000	0

Note: This table is for projects (like CS) where the final occupant count is not available. If your project's occupancy factors are not listed above, you can use a comparable building to show the average gross sf per occupant for your building's use.

2. **Important resources and further study materials you can download for <u>free</u>**

Energy Performance of LEED® for New Construction Buildings: Final Report, by Cathy Turner and Mark Frankel (2008):
http://www.gbci.org/Files/References/Energy-Performance-of-LEED-for-New-Construction-Buildings-Final-Report.pdf

Foundations of the Leadership in Energy and Environmental Design Environmental Rating System: A Tool for Market Transformation (LEED Steering Committee, 2006):
http://www.gbci.org/Files/References/Foundations-of-the-Leadership-in-Energy-and-Environmental-Design-Environmental-Rating-System-A-Tool-for-Market-Transformation.pdf

AIA Integrated Project Delivery: A Guide (www.aia.org):
http://www.aia.org/contractdocs/AIAS077630

Review of ANSI/ASHRAE Standard 62.1-2007: Ventilation for Acceptable Indoor Air Quality, by Brian Kareis:
http://www.workplace-hygiene.com/articles/ANSI-ASHRAE-3.html

Best Practices of ISO-14021: Self-Declared Environmental Claims, by Kun-Mo Lee and Haruo Uehara (2003):

http://www.ecodesign-company.com/documents/BestPracticeISO14021.pdf

Bureau of Labor Statistics (www.bls.gov)

International Code Council (www.iccsafe.org)

Americans with Disabilities Act (ADA): Standards for Accessible Design (www.ada.gov):
http://www.ada.gov/stdspdf.htm

GSA 2003 Facilities Standards (General Services Administration, 2003):
http://www.gbci.org/Files/References/GSA-2003-facilities-standards.pdf

Guide to Purchasing Green Power (Environmental Protection Agency, 2004):
http://www.gbci.org/Files/References/Guide-to-Purchasing-Green-Power.pdf

USGBC Definitions:
https://www.usgbc.org/ShowFile.aspx?DocumentID=5744

3. **Annotated bibliography**

   Chen, Gang. **LEED GA MOCK EXAMS:** *Questions, Answers, and Explanations: A Must-Have for the LEED Green Associate Exam, Green Building LEED Certification, and Sustainability*. ArchiteG, Inc, the latest version. This is a companion to *LEED Green Associate Exam Guide (LEED GA)*. It includes 200 questions, answers, and explanation, and is very close to the real LEED Green Associate Exam.

4. **Valuable websites and links**

a. The official website for the US Green Building Council (USGBC) is at the following link:
   http://www.usgbc.org/

   Pay special attention to the purpose of LEED Online, LEED project registration, LEED certification content, LEED reference guide introductions, LEED rating systems, and checklists.

   You can download or purchase the following useful documents from the USGBC or GBCI Web site: Latest and official LEED exam candidate handbooks including an exam content outline and sample questions:
   http://www.gbci.org/main-nav/professional-credentials/resources/candidate-handbooks.aspx

   LEED Reference Guides, **Guide to certification**, and various versions of LEED Green Building Rating Systems and Project Checklist:
   http://www.usgbc.org/projecttools

   Read the **Guide to certification** at least three times, because it is VERY important, and it tells you which LEED system to use.

USGBC issue LEED Addenda for various LEED Green Building Rating **Systems** and **reference guides** on a quarterly basis. **Make sure you download the latest LEED Addenda** related to your exam and read them at least three times. See link below for detailed information:
http://www.usgbc.org/addenda

b. Natural Resources Defense Council:
   http://www.nrdc.org/

c. Environmental Construction + Design - Green Book (Offers print magazine and online environmental products and services resources guide):
   http://www.edcmag.com/greenbook

d. Cool Roof Rating Council Web site:
   http://www.coolroofs.org

5. Important items covered by the second edition of *Green Building and LEED Core Concepts Guide*

Starting on December 1, 2011, GBCI will begin to draw LEED Green Associate Exam questions from the second edition of *Green Building and LEED Core Concepts Guide*. The following are some "new" and important items covered by this edition:

adaptive reuse: Designing and constructing a building to accommodate a future use that is different from its original use.

biomimicry: Learning from nature and designing systems using principles that have been tested in nature for millions of years.

carbon overlay: LEED credit weighting based on each credit's impact on reducing carbon footprint.

charrettes: Intensive (design) workshops.

cradle to cradle: A method where materials are used in a closed system and generate no waste.

cradle to grave: A process that examines materials from their point of extraction to disposal.

closed system: There is no "away." Everything goes somewhere within the system, the waste generated by a process becomes the "food" of another process. Nature is a closed system.

embodied energy: The total energy consumed by extracting, harvesting, manufacturing, transporting, installing, and using a material through its entire life cycle.

ENERGY STAR's Portfolio Manager: An online management tool for tracking and evaluating water and energy use. An ENERGY STAR Portfolio Manager score of 50 means a building is at national average energy use level for its category. A score higher than 50 means a building is more energy efficient than the national average energy use level for its category. The higher the score, the better.

evapotranspiration: Loss of water due to evaporation.

externalities: Benefits or costs that are NOT part of a transaction.

feedback loop: Information flows within a system that allows the system to adjust itself. A thermostat or melting snow is an example of negative feedback loop. Population growth, heat island effect, or climate change is a positive feedback loop. Positive feedback loop can create chaos in a system.

International Green Construction Code (IGCC): A national model green building code published by International Code Council (ICC).

integrated process: Emphasizes communications and interactions among stakeholders throughout the life of a project. Integrated process is a holistic decision making process based on systems thinking and life cycle approach.

interative process: A repetitive and circular process that helps a team to define goals and check ideas against these goals.

Integrated Pest Management (IPM): A sustainable approach to pest management.

LEED interpretations: Precedent-setting (project credit interpretation) rulings. A project team can opt into the LEED interpretation process when submitting an inquiry to GBCI.

leverage points: Places where a small interventions can generate big changes.

life cycle approach: Looking at a product or building through its entire life cycle.

life cycle assessment (LCA): Use life cycle thinking in environmental issues.

life cycle costing: Looking at the cost of purchasing and operating a building or product, and the relative savings.

low impact development (LID): A land development approach mimicking natural systems and managing storm water as close to the source as possible.

"Net-Zero": A project, which doesn't use any more resources than what it can produce. Similar concepts include carbon neutrality and water balance.

negative feedback loop: A signal for the system to stop changes when a response is not needed anymore.

open system: Resources are brought from the outside, consumed, and then disposed of as waste to the outside.

permaculture: Designing human habitats and agriculture systems based on models and relationships found in nature.

positive feedback loop: A stimulus causes an effect and encourages the loop to produce more of this effect.

Prius effect: Provides real time feedback of energy use so that users can adjust behaviors to save energy.

Project CIRs: LEED credit interpretation rulings for specific project circumstances.

retrocommissioning: A building tune-up that restores efficiency and improves performance.

regenerative: Regenerative buildings and communities evolve with living systems and help to renew resources and life. Regenerative projects generate electricity and sell the excess back to the grid, as well as return water to nature, which is cleaner than it was before use.

systems thinking: In a system, each component affects many other components. They are all related to each other.

Wingspread Principles on the US Response to Global Warming: A set of principles signed by organizations and individuals to express their commitment to address global warming. It calls for 60% to 80% reduction of green house gas emission by midcentury (based on 1990 levels).

Back page promotion:

Other useful books written by Gang Chen:

1. **LEED Exam Guide series.** Refer to the links below:
   http://www.GreenExamEducation.com
   http://www.ArchiteG.com

2. **Building Construction:** Project Management, Construction Administration, Drawings, Specs, Detailing Tips, Schedules, Checklists, and Secrets Others Don't Tell You (Architectural Practice Simplified, 2nd edition)

3. *Planting Design Illustrated: A Must-Have for Landscape Architecture: A Holistic Garden Design Guide with Architectural and Horticultural Insight, and Ideas from Famous Gardens in Major Civilizations* (2nd edition):
   http://www.GreenExamEducation.com

Note: Other books in the LEED Exam Guide series are currently in production. One book will eventually be produced for each of the LEED exams. The series includes:

LEED Green Associate Exam Guide (LEED GA): *Comprehensive Study Materials, Sample Questions, Mock Exam, Green Building LEED Certification, and Sustainability (LEED v3.0)*, Book 2, LEED Exam Guide series, ArchiteG.com (latest edition)

LEED BD&C EXAM GUIDE: *A Must-Have for the LEED AP BD+C Exam: Comprehensive Study Materials, Sample Questions, Mock Exam, Green Building Design and Construction, LEED Certification, and Sustainability*, Book 3, LEED Exam Guide series, ArchiteG.com (latest edition)

LEED ID&C EXAM GUIDE: *A Must-Have for the LEED AP ID+C Exam: Comprehensive Study Materials, Sample Questions, Mock Exam, Green Interior Design and Construction, LEED Certification, and Sustainability*, Book 4, LEED Exam Guide series, ArchiteG.com (latest edition)

LEED O&M EXAM GUIDE: *A Must-Have for the LEED AP O+M Exam: Comprehensive Study Materials, Sample Questions, Mock Exam, Green Building Operations and Maintenance, LEED Certification, and Sustainability*, Book 5, LEED Exam Guide series, ArchiteG.com (latest edition)

LEED HOMES EXAM GUIDE: *A Must-Have for the LEED AP+ Homes Exam: Comprehensive Study Materials, Sample Questions, Mock Exam, Green Building LEED Certification, and Sustainability*, Book 6, LEED Exam Guide series, ArchiteG.com (latest edition)

LEED ND EXAM GUIDE: *A Must-Have for the LEED AP+ Neighborhood Development Exam: Comprehensive Study Materials, Sample Questions, Mock Exam, Green Building LEED Certification, and Sustainability*, Book 7, LEED Exam Guide series, ArchiteG.com (latest edition)

LEED GA MOCK EXAMS: *Questions, Answers, and Explanations: A Must-Have for the LEED Green Associate Exam, Green Building LEED Certification, and Sustainability*, Book 8, LEED Exam Guide series, ArchiteG.com (latest edition)

LEED O&M MOCK EXAMS: *Questions, Answers, and Explanations: A Must-Have for the LEED O&M Exam, Green Building LEED Certification, and Sustainability*, Book 9, LEED Exam Guide series, ArchiteG.com (latest edition)

How to order these books:
You can order the books listed above at:
http://www.GreenExamEducation.com

OR get free tips and information at
http://www.GeeForums.com

Following are some detailed descriptions of each text:

LEED Exam Guide series
Comprehensive Study Materials, Sample Questions, Mock Exam, Building LEED Certification, and Going Green

LEED (Leadership in Energy and Environmental Design) is the most important trend in development and is revolutionizing the construction industry. It has gained tremendous momentum and has a profound impact on our environment. From the LEED Exam Guide series, you will learn how to:

1. Pass the LEED Green Associate Exam and various other LEED AP+ exams (each book will help you with a specific LEED exam).

2. Register and certify a building for LEED certification.

3. Understand the intent of each LEED prerequisite and credit.

4. Calculate points for a LEED credit.

5. Identify the responsible party for each prerequisite and credit.
6. Earn extra credits (exemplary performance) for LEED.

7. Implement the local codes and building standards for prerequisites and credits.

8. Receive points for categories not yet clearly defined by the USGBC.

There is currently NO official GBCI book on any of the LEED exams, and most of the existing books on LEED and LEED AP+ are too expensive and too complicated to be practical or helpful. The pocket guides in the LEED Exam Guide series fill in the blanks, demystify LEED, and uncover the tips, codes, and jargon for LEED, as well as the true meaning of "going green." They will set up a solid foundation and fundamental framework of LEED for you. Each book in the LEED Exam Guide series covers every aspect of one or more specific LEED rating system in plain and concise language, and makes this information understandable to anyone.

These pocket guides are small and easy to carry to read when time permits. They are indispensable books for everyone: administrators; developers; contractors; architects; landscape architects; civil, mechanical, electrical, and plumbing engineers; interns; drafters; designers; and other design professionals.

Why is the LEED Exam Guide series needed?

A number of books are available that you can use to prepare for the LEED exams. Consider the following:

1. USGBC reference guides. You need to select the correct version of the reference guide for your exam.

The USGBC reference guides are comprehensive, but they give too much information. For example, *The LEED Reference Guide for Building Design and Construction (v4 for BD&C)* has approximately 817 oversized pages. Many of the calculations in the books are too detailed for the exam. The books are also expensive (approximately $250 each, so most people may not buy them for their personal use, but instead, will seek to share an office copy).

Reading a reference guide from cover to cover is good if you have the time. The problem is that very few people actually have the time to read the whole reference guide. Even if you do read the whole guide, you may not remember the important issues required to pass the LEED exam. You need to reread the material several times before you can remember much of it.

Reading a reference guide from cover to cover without a guidebook is a difficult and inefficient way of preparing for the LEED exams, because you do NOT know what USGBC and GBCI are looking for in the exam.

2. The USGBC workshops and related handouts are concise, but they do not cover extra credits (exemplary performance). The workshops are expensive, costing approximately $450 each.

3. Various books published by third parties are available. However, most of them are not very helpful.

There are many books on LEED, but not all are useful.

Each book in the LEED Exam Guide series will fill in the blanks and become a valuable, reliable source.

a. They will give you more information for your money. Each of the books in the LEED Exam Guide series provides more information than the related USGBC workshops.

b. They are exam-oriented and more effective than the USGBC reference guides.

c. They are better than most, if not all, of the other third-party books. They give you comprehensive study materials, sample questions and answers, mock exams and answers, and critical information on building LEED certification and going green. Other third-party books only provide a fraction of this information.

d. They are comprehensive yet concise, small, and easy to carry around. You can read them whenever you have a few spare minutes.

e. They are great timesavers. I have highlighted the important information that you need to understand and MEMORIZE. I also make some acronyms and short sentences to help you easily remember the credit names.

You should devote about 1 to 2 weeks of full-time study to pass each of the LEED exams. I have met people who have spent only 40 hours of study time and passed the exams.

You can find sample texts and other information about the LEED Exam Guide series listed under the customer discussion section for each available book.

What others are saying about *LEED GA Mock Exams* (Book 8, LEED Exam Guide series):

"Great news, I passed!!! As an educator and business professional I would absolutely recommend this book to anyone looking to take and pass the LEED Green Associate exam on the first attempt."
—**Luke Ferland**

"Elite runners will examine a course, running it before they race it...This book is designed to concentrate on increasing the intensity of your study efforts, examine the course, and run it before you race it..."
—**Howard Patrick (Pat) Barry, AIA NCARB**

"Like many similar test prep guides, Mr. Chen cites the resources that will be useful to study. But he goes beyond this and differentiates which ones must memorize and those you must be at least familiar with."
—**NPacella**

"Read *LEED GA Mock Exams* before you start studying other resource materials. It will serve to bring your attention to the information that you are most likely to be asked on the exam as you come across it in your studying."
—**Mike Kwon**

"I found these exams to be quite tougher compared to the others I took a look at, which is good as it made me prepare for the worst I would definitely recommend using these mock exams. I ultimately passed with 181..."
—**Swankysenor**

Building Construction

Project Management, Construction Administration, Drawings, Specs, Detailing Tips, Schedules, Checklists, and Secrets Others Don't Tell You (Architectural Practice Simplified, 2nd edition)

Learn the Tips, Become One of Those Who Know Building Construction and Architectural Practice, and Thrive!

For architectural practice and building design and construction industry, there are two kinds of people: those who know, and those who don't. The tips of building design and construction and project management have been undercover—until now.

Most of the existing books on building construction and architectural practice are too expensive, too complicated, and too long to be practical and helpful. This book simplifies the process to make it easier to understand and uncovers the tips of building design and construction and project management. It sets up a solid foundation and fundamental framework for this field. It covers every aspect of building construction and architectural practice in plain and concise language and introduces it to all people. Through practical case studies, it demonstrates the efficient and proper ways to handle various issues and problems in architectural practice and building design and construction industry.

It is for ordinary people and aspiring young architects as well as seasoned professionals in the construction industry. For ordinary people, it uncovers the tips of building construction; for aspiring architects, it works as a construction industry survival guide and a guidebook to shorten the process in mastering architectural practice and climbing up the professional ladder; for seasoned architects, it has many checklists to refresh their memory. It is an indispensable reference book for ordinary people, architectural students, interns, drafters, designers, seasoned architects, engineers, construction administrators, superintendents, construction managers, contractors, and developers.

You will learn:
1. How to develop your business and work with your client.
2. The entire process of building design and construction, including programming, entitlement, schematic design, design development, construction documents, bidding, and construction administration.
3. How to coordinate with governing agencies, including a county's health department and a city's planning, building, fire, public works departments, etc.
4. How to coordinate with your consultants, including soils, civil, structural, electrical, mechanical, plumbing engineers, landscape architects, etc.
5. How to create and use your own checklists to do quality control of your construction documents.
6. How to use various logs (i.e., RFI log, submittal log, field visit log, etc.) and lists (contact list, document control list, distribution list, etc.) to organize and simplify your work.
7. How to respond to RFI, issue CCDs, review change orders, submittals, etc.
8. How to make your architectural practice a profitable and successful business.

Planting Design Illustrated
A Must-Have for Landscape Architecture: A Holistic Garden Design Guide with Architectural and Horticultural Insight, and Ideas from Famous Gardens in Major Civilizations

One of the most significant books on landscaping!

This is one of the most comprehensive books on planting design. It fills in the blanks of the field and introduces poetry, painting, and symbolism into planting design. It covers in detail the two major systems of planting design: formal planting design and naturalistic planting design. It has numerous line drawings and photos to illustrate the planting design concepts and principles. Through in-depth discussions of historical precedents and practical case studies, it uncovers the fundamental design principles and concepts, as well as the underpinning philosophy for planting design. It is an indispensable reference book for landscape architecture students, designers, architects, urban planners, and ordinary garden lovers.

What Others Are Saying About *Planting Design Illustrated* ...

"I found this book to be absolutely fascinating. You will need to concentrate while reading it, but the effort will be well worth your time."
—**Bobbie Schwartz, former president of APLD (Association of Professional Landscape Designers) and author of *The Design Puzzle: Putting the Pieces Together*.**

"This is a book that you have to read, and it is more than well worth your time. Gang Chen takes you well beyond what you will learn in other books about basic principles like color, texture, and mass."
—**Jane Berger, editor & publisher of gardendesignonline**

"As a longtime consumer of gardening books, I am impressed with Gang Chen's inclusion of new information on planting design theory for Chinese and Japanese gardens. Many gardening books discuss the beauty of Japanese gardens, and a few discuss the unique charms of Chinese gardens, but this one explains how Japanese and Chinese history, as well as geography and artistic traditions, bear on the development of each country's style. The material on traditional Western garden planting is thorough and inspiring, too. *Planting Design Illustrated* definitely rewards repeated reading and study. Any garden designer will read it with profit."
—**Jan Whitner, editor of the *Washington Park Arboretum Bulletin***

"Enhanced with an annotated bibliography and informative appendices, *Planting Design Illustrated* offers an especially "reader friendly" and practical guide that makes it a very strongly recommended addition to personal, professional, academic, and community library gardening & landscaping reference collection and supplemental reading list."
—**Midwest Book Review**

"Where to start? *Planting Design Illustrated* is, above all, fascinating and refreshing! Not something the lay reader encounters every day, the book presents an unlikely topic in an easily digestible, easy-to-follow way. It is superbly organized with a comprehensive table of contents, bibliography, and appendices. The writing, though expertly informative, maintains its accessibility throughout and is a joy to read. The detailed and beautiful illustrations expanding on the concepts presented were my favorite portion. One of the finest books I've encountered in this contest in the past 5 years."
—Writer's Digest 16th Annual International Self-Published Book Awards Judge's Commentary

"The work in my view has incredible application to planting design generally and a system approach to what is a very difficult subject to teach, at least in my experience. Also featured is a very beautiful philosophy of garden design principles bordering poetry. It's my strong conviction that this work needs to see the light of day by being published for the use of professionals, students & garden enthusiasts."
—Donald C. Brinkerhoff, FASLA, chairman and CEO of Lifescapes International, Inc.

Index

3016 rule, 13, 25, 26, 27
Active (Mechanical) Ventilation, 132
Actual materials cost, 114
actual measurement, 143
actual performance, 86, 100, 143
albedo, 42
Angoff Method, 34
ANSI, 49, 109
ASHRAE, 41, 99, 109, 133, 139, 154, 155, 158, 159, 160, 171
association factors, 38
ASTM, 72
BD&C, 31, 33, 35, 36, 53
BECx, 102
blowdown, 89
BMPs, 70, 133
BUG, 69, 78, 155, 160
CARB, 138, 156, 161
Carbon offsets, 108
CBE, 145
CFC, 42, 101, 107, 157, 162
CGP, 71
CHP, 109
CI, 31, 35, 44, 49
CIR, 42, 46, 48, 49, 52, 167
CMP, 31, 34
Commissioning, 51, 95, 96
computer simulation, 143
cradle to cradle, 112, 116, 173
cradle to grave, 112, 173
CRI, 161
CSRs, 123, 155, 160
curtailment event, 105
CWM, 118
Cx, 98, 102
CxA, 95, 98, 102
Default materials cost, 114
demand response program, 95
dematerialization, 120
densely occupied space, 131
detailed design analysis and input, 143
distracters, 23, 158, 159, 160, 161
DR, 51, 105
energy modelers, 99
EPA, 82, 92
EPAct, 82
EPDs, 122
ESA, 72
ESC, 71
ETS, 132, 135, 154, 159
eutrophication, 69, 120
field-based research, 141
FTE, 42, 90, 91, 156, 161, 171
GBCI, 19, 23, 24, 31, 32, 33, 34, 35, 49, 52, 53, 127, 163
GI, 76
Green Building and LEED Core Concepts Guide, 18, 19, 173
Green Rater, 47
Green Seal, 41, 116, 156, 161
Green-e, 42
GREENGUARD, 161
GWP, 107
halons, 42
HCFC, 42, 101, 107, 157, 162
heat islands, 77
HET, 92, 93
HEU, 92, 93
High-efficiency, 93
IAQ, 132, 134, 137, 139, 140, 141, 154, 158, 159
ID&C, 31, 32, 35, 36
IES, 78, 142, 160
IESNA, 109, 155, 160
Impact Categories, 38
Incentive programs, 105
individual occupant space, 131
IPC, 14, 56, 73, 76, 92, 98, 99
ISO, 49, 112, 122, 172
LCA, 112, 120, 174
LEED AP+, 19, 31, 32, 33
LEED for Health Care, 36
LEED for Homes, 47, 167
LEED for Retails, 36, 37
LEED Online, 42, 43, 46, 52, 53, 167, 172
LEED v4, 14, 17, 19, 37, 39, 41, 44, 45, 46, 47, 50, 111, 113, 145, 166
LEED-CS, 49
LEED-EB, 31, 32, 35, 37
LEED-NC, 11, 31, 35, 43
LEED-ND, 37
LID, 76, 174

life-cycle approach, 111
Location Valuation Factor, 114
LWR, 84
main credit categories, 52, 53
MasterFormat, 113
MBCx, 102
MEP, 113, 114
MERV, 139, 154, 159
Mnemonic, 13, 23, 25, 26, 27, 35, 51, 82, 96, 115, 116, 127, 132
MPR, 49, 112
non-densely occupied space, 131
non-regularly occupied, 130, 131
NRDC, 41
O&M, 31, 32, 35, 37, 44, 46, 49, 51
occupancy type, 90
ODP, 107, 157, 162
Passive (Natural) Ventilation, 132
PBTs, 124
performance approach, 99, 158
PMV, 141
POPs, 124
PPD, 141
precertification, 48
prescriptive approach, 99, 158
Process energy, 109
process of memorization, 24
Prometric, 163

psychometricians, 34
RECs, 108
Regularly occupied spaces, 130
Regulated (non-process) energy, 109
SCAQMD, 138, 155, 156, 159, 161
SCM, 138, 156, 161
shared multi-occupant space, 131
simulated daylight analysis, 143
six primary factors, 141, 158
SMACNA, 41, 139, 155, 159
solid waste management hierarchy, 111
Source reduction, 111, 118
SR, 69, 77
SRI, 42, 69, 77
stakeholders, 33
subject matter experts, 34
synergy, 41
three forms of light pollution, 78
Tiered demand electricity pricing, 105
triple bottom lines, 36
UL, 161
UPC, 92
VOCs, 138, 140
VRF, 107
WaterSense, 82, 92
weighting process, 39
whole building energy simulation, 158

CPSIA information can be obtained
at www.ICGtesting.com
Printed in the USA
LVHW061615280819
629260LV00009B/398/P